CONSTRUCTION PROCESSES
AND
MATERIALS

CONSTRUCTION PROCESSES AND MATERIALS

D. E. WARLAND
A.I.O.B.

HODDER AND STOUGHTON
LONDON SYDNEY AUCKLAND TORONTO

First printed 1965
Reprinted 1976

ISBN 0 340 04582 5

Printed in Great Britain for Hodder and Stoughton Educational, a division of Hodder and Stoughton Ltd, London, by J. W. Arrowsmith Ltd, Bristol.

Preface

FOR a number of years my father's books on Building Construction have provided text books for the National Certificate and Diploma courses in Building. With the revision of the National Certificate courses coinciding with the completion of an edition of these books, we considered it advisable to bring them into line with the new schemes. This book has been written, therefore, primarily for the subject of Construction for the two years of the General Course in Construction, but will be useful for other courses offering elementary Construction, such as the first year of the Construction Technician course. The book is intended as a text book to augment the normal teaching of the subject and covers the basic construction of a bungalow, being related as far as possible to the bungalow illustrated at the beginning of the book.

In conclusion, I would pay tribute to my father, in this first book to be published since his recent death, for his inspiration and encouragement over the many years we have been together in Technical Education. I would also thank all those who have given help in the writing of this book, and Messrs Clarke Bros of Leamington Spa for permission to use the drawing of the bungalow.

D. E. W.

Contents

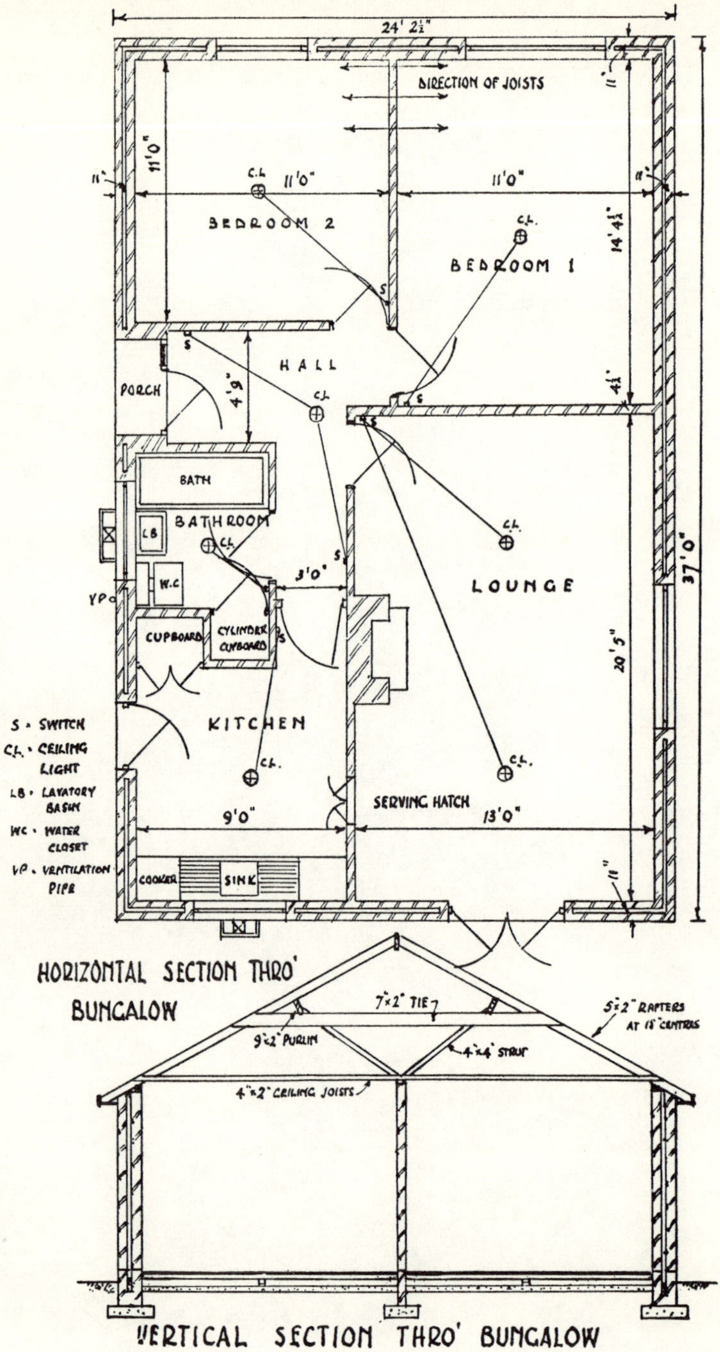
24' 2½"
DIRECTION OF JOISTS
11'0"
11'0"
11'0"
11"
11"
11"
14' 4½"
BEDROOM 2
BEDROOM 1
C.L.
S
HALL
PORCH
4' 9"
4½"
BATH
BATHROOM
LB
W.C
VP
3'0"
CUPBOARD
CYLINDER CUPBOARD
LOUNGE
37' 0"
20' 5"
KITCHEN
SERVING HATCH
9'0"
13'0"
COOKER
SINK
S = SWITCH
C.L. = CEILING LIGHT
LB = LAVATORY BASIN
WC = WATER CLOSET
VP = VENTILATION PIPE
HORIZONTAL SECTION THRO' BUNGALOW
7"×2" TIE
5"×2" RAFTERS AT 15" CENTRES
9"×2" PURLIN
4"×4" STRUT
4"×2" CEILING JOISTS
VERTICAL SECTION THRO' BUNGALOW

Figure 1

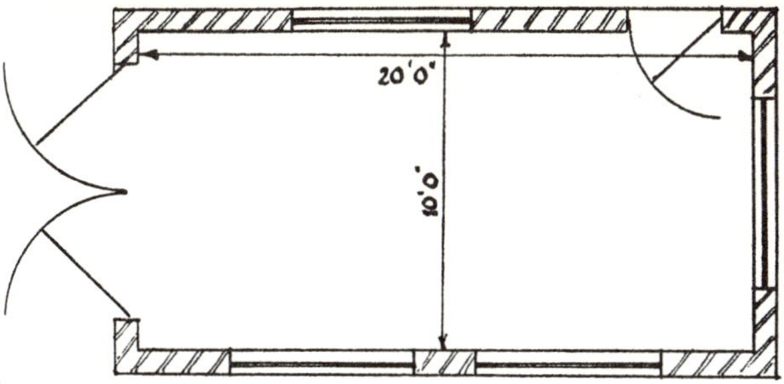

HORIZONTAL SECTION THRO' GARAGE

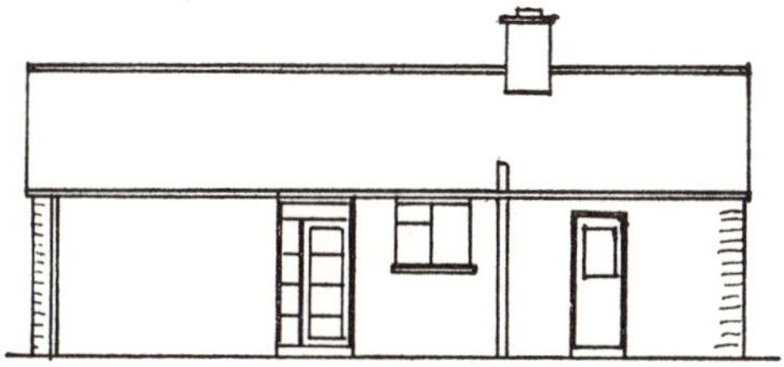

FRONT ELEVATION OF BUNGALOW

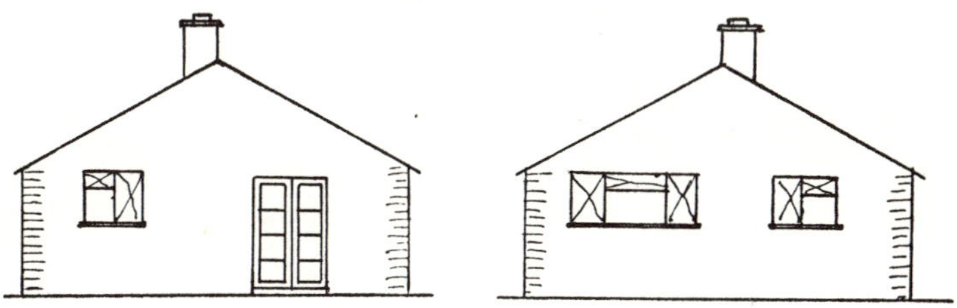

SIDE ELEVATIONS OF BUNGALOW

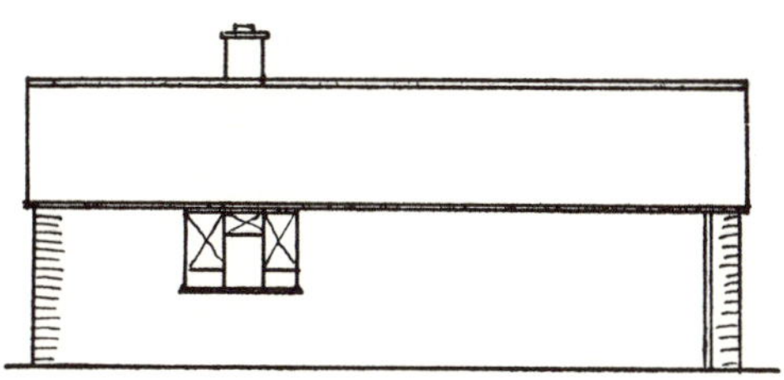

BACK ELEVATION

Figure 2

1 Preliminary Site Work

Choice of Site

In most instances the choice of a site is not the builder's but the client's. In estate development, however, the builder will purchase and develop the site according to the local housing demands.

When the client is able to choose a site for a domestic building, the aspect (i.e. the direction the house faces), the nearness of other property, the Local Authority's future planning and designation of the area, particularly if open spaces exist adjacent or near the proposed site, should be considered.

The site having been chosen and purchased the client will employ an architect to design the house to suit the site conditions.

Access to Site

The site having been handed over to the builder, the best method of access must be determined to suit the position and condition of the site.

When the site is adjacent to an existing road the pavement must be protected with sleepers, and if the ground is soft within the site, sleepers or hardcore should be laid to enable lorries to enter and leave the site without being stuck.

When the site is an estate, or over a large area, the permanent roads should be constructed in concrete, or with the hardcore base, to allow traffic to move easily.

Temporary Services

It is often advantageous to the contractor to have the permanent services connected, particularly water and electricity, and bring them on the site, from which temporary supplies can be connected later. Water will be connected temporarily to a stand pipe near the mixer. It should be adequately protected against damage and frost.

Shelter for Workmen and Materials

The builder should provide a hut for the workmen so that meals and tea breaks can be taken in the dry and also be available for storage of clothing not required for work during the day, and protective clothing at night. This is a requirement of the Working Rule Agreement and the Factory Act. Latrines must also be provided.

Materials, such as cement, etc., should be protected from weather by storing in a shed, or in the case of metal windows, facing bricks, etc., stacked in a suitable position on the site, where they will not be liable to damage, and are adequately protected. Ironmongery, sanitary fittings, etc., should be kept in a locked shed, or in a locked room of the building being erected.

Site Work

The site should be cleared of all bushes, shrubs, hedges, trees, etc., which are on the building area. The roots should be grubbed up and completely removed. Trees which would be near the building and which may affect the foundations by root growth must also be removed. Care must be taken in removing trees that there is no infringement of any tree preservation order.

Before any building is erected it is essential that the area to be occupied by the building has the vegetable soil removed by excavating, since the organic content may be injurious to the construction and to living conditions within the structure. The depth of the excavation will vary according to the geographical strata and formation, normally it will be about 4″ to 6″ deep, but may be as deep as 10″.

The vegetable soil should be placed on one side during building operations and spread and levelled over areas after completion to provide gardens, etc., or it should be taken off the site completely. It should never be used for backfilling or making up levels under the building for the reasons given above.

Setting Out

It is important that the structure is built in the correct position as shown on the architect's drawings, as in many cases deviation may infringe the building line or other legal requirements.

The *building line* is an imaginary line, determined by the Local Authority, set back from the roadway, in front of which buildings must not encroach.

The initial setting out of the building will be carried out by the foreman or the surveyor and will be carefully checked. Fixed positions on other nearby buildings may be used to set out and check the position of the new structure.

To set out a building the front line should be determined, which in the case of a domestic building can often be related to the road-

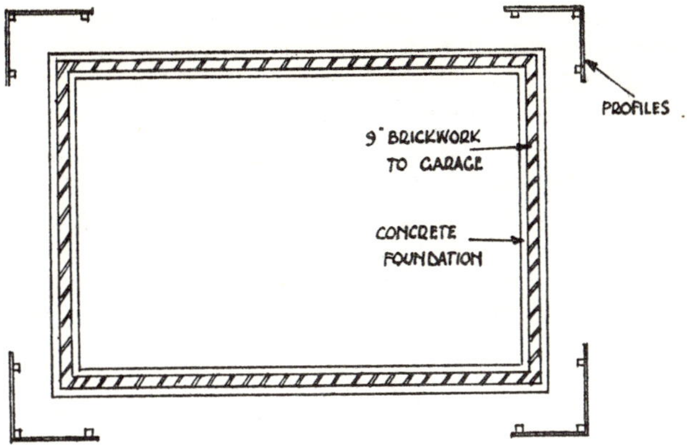

(a) PROFILES TO GARAGE FOUNDATION

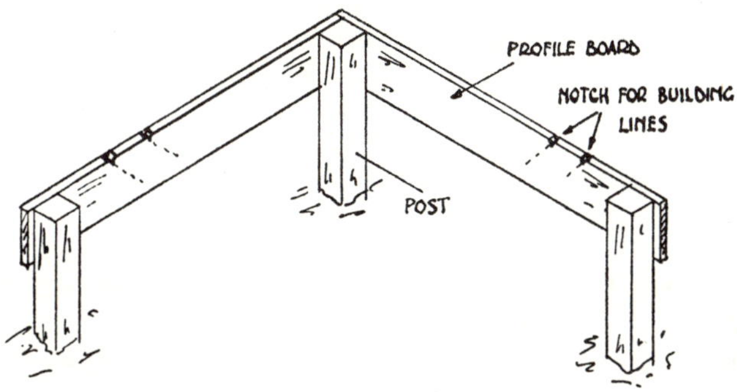

(b) SKETCH OF PROFILE

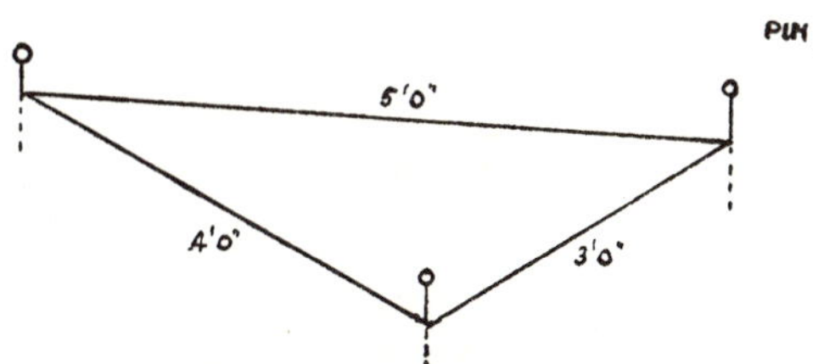

(c) SKETCH OF METHOD OF SETTING-OUT
A RIGHT ANGLE

Figure 3

way. Having marked its position and length the return walls at each end can be marked out. A right angle can be obtained with a tape or measuring rod based on the 3:4:5 rule, viz. a triangle with sides in this proportion will give a right angle triangle as illustrated in Figure 3. The length of the return can be measured and the back of the building marked out. To check the accuracy of the rectangle formed, the diagonals should be found to be equal in length.

The position of the building having been established and pegged, *profiles* will be erected. Profiles must be placed in positions where they will not impede the work, their function being to mark the position of the trenches for the foundation and the brickwork for the external walls. A profile will be placed at each corner and at any return in the wall, as shown in Figure 3.

2 Foundations

THE ground below the vegetable soil is known as the subsoil. The type of subsoil which will be encountered will range from solid rock to waterlogged ground. Special precautions and methods will be necessary in dealing with the extremes, but most building work will be in the middle range of subsoil from good firm subsoil to soft subsoil.

The purpose of foundations is to provide suitable support and stability for the structure according to the type of subsoil encountered.

Trenches

Depth. For the range of average subsoils a foundation trench will be necessary for two main reasons:

1. The subsoil will need to be of sufficient strength (termed 'bearing capacity') to support the load of the structure to be erected. It will be necessary, therefore, to excavate through the subsoil until it is of sufficient strength. The depth will vary according to the subsoil and the Local Authority will require to inspect the trench to ensure that a good bottom has been reached.
2. The action of weather on the subsoil will also affect the depth to which the foundations should be taken. The change in the climatic condition over the year will cause the subsoil to shrink and swell and during cold weather frost will have an action. Gravel subsoils are not affected greatly by moisture movement or frost, and a trench depth of 1′ 6″ to 2′ 0″ will be sufficient. Clay on the other hand will move considerably under moisture movement and the depth of trench should be increased to up to 4′ 0″.

Width. The width of trenches for foundations will be determined by two considerations:

1. The width necessary for the concrete foundations. For this width see page 10.

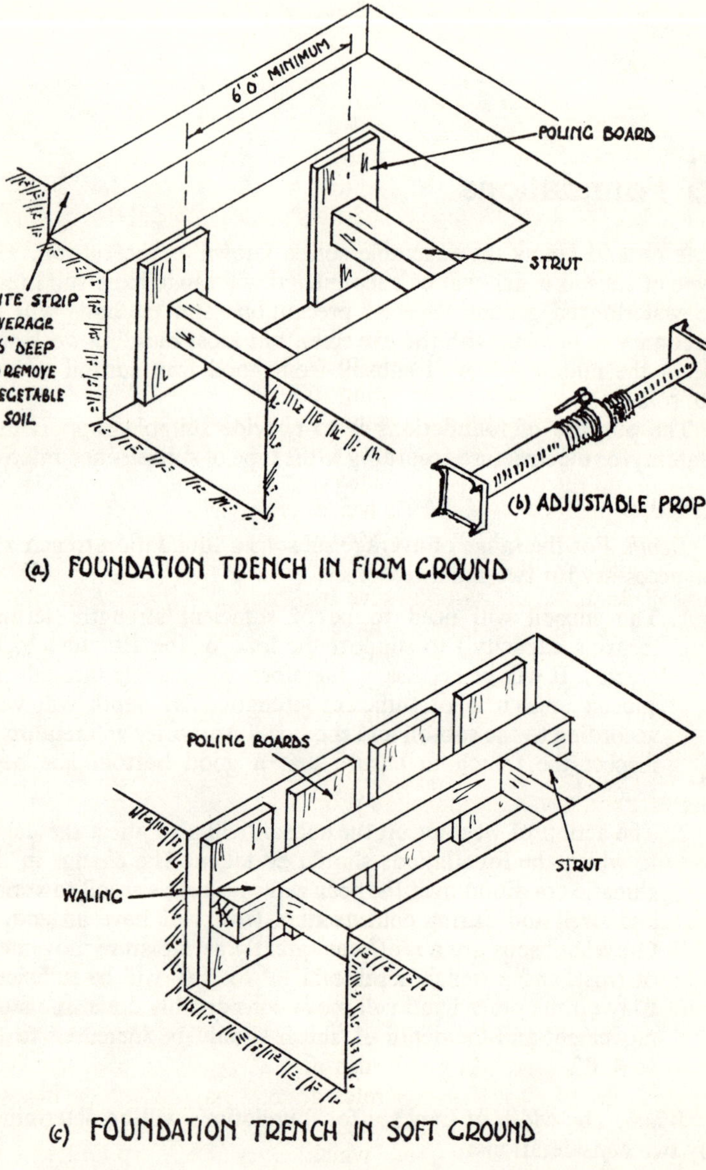

(a) FOUNDATION TRENCH IN FIRM GROUND

(c) FOUNDATION TRENCH IN SOFT GROUND

Figure 4

2. The width necessary for a man to work in the trench to lay the foundations and build the brickwork up to ground level.

Timbering

The responsibility for the provision of timbering to the sides of trenches is the builder's. The amount of timbering will depend upon (*a*) the type of subsoil to be supported, (*b*) the length of time the trench is to remain open, (*c*) the time of year or climatic conditions prevailing when the trench is excavated and (*d*) the depth of excavation. It is advisable to have some form of timbering or support as subsoils can be treacherous particularly in the clay range. The builder should err on the side of safety in trench work rather than gambling with the lives and limbs of those who are required to work in the trench. A man need only to be buried up to his waist to be killed. Apart from death and injury it will be additional cost to the builder to re-excavate and renew damaged work in the trench should the sides collapse. The need for the builder and his employees to carry out the safety requirements cannot be emphasised too much as the Construction Industry has a very high accident rate and everyone in the Industry must become more safety conscious.

As a guide for the type of timbering which can be used for various subsoil, Figures 4(a) and 4(c) give two types of subsoil with appropriate timbering. The drawings have been based on the use of timber, but there are other methods and materials which can be used with equal effectiveness and greater ease of erection. Metal trough sections can replace the more usual poling boards, and metal adjustable props can replace the timber struts. The advantages of the adjustable prop are the elimination of wedges and also easier adjustment. Figure 4(b) shows a sketch of an adjustable prop.

Concrete

Concrete as a material has been known since before Roman times, but it is only since the introduction of Portland cement in 1824 that it has been used widely as a structural material.

Concrete is composed of coarse and fine aggregate mixed with cement.

Coarse aggregate will be gravel, shingle or ballast. It should be of strong material not likely to crush and be clean, i.e. free from silt or clay. For very strong concretes granite may be used, while for lightweight concretes, foamed slag, pumice, etc., may be used.

Fine aggregate will be sand, which should be sharp and washed clean.

Aggregate should be from pit or river sources and not from the sea shore. Sea shore aggregates are usually round and contain salt. The salt will not affect the strength of the concrete, but will cause efflorescence on the face of the structure and disfigure it.

Cement

The most common is *ordinary Portland cement*. It is manufactured from chalk and clay in various parts of the country, where these materials are found. Clay is excavated, combined with chalk or limestone and ground with water to form a slurry. The slurry having been screened is passed into a storage tank and agitated in preparation for passing into the kiln. The kiln is a long inclined rotating cylinder, with a refractory liming and is fired by pulverised coal. The slurry is passed into the higher end of the kiln and flows down as it revolves. The kiln dries the slurry as it flows, the water being evaporated and the temperature raised to about 1500°C, fusing the materials to a clinker. The clinker is cooled and ground to a fine powder. A small amount of gypsum is added to retard the setting of the cement.

Materials	*Process*
Limestone or Chalk Clay Water	Wet Grinding Mill
	Slurry
	Screen
	Storage Tank
	Kiln Feeder
Pulverised Coal	Rotary Kiln
	Clinker Coolers
	Clinker Storage Hoppers
Gypsum	Clinker Grinding Mills
	Cement Silo
	Despatch

Other cements which are commonly used are:

Rapid-hardening Portland cement hardens more quickly than, but is similar to, ordinary Portland cement, except that it is more finely ground and has a slight difference in the chemical composition.

High-alumina cement is not of the same composition as ordinary Portland cement consisting of chalk and bauxite. It has the advantage of generating heat during the setting period which makes it suitable for concreting during cold weather. It hardens quickly but sets slowly and is more resistant to sulphates and sea-water than ordinary Portland cement.

Proportions of Mix

Concrete may be proportioned by volume or by weight; the latter is gradually superseding the former. Typical mixes are given in the table below as set out in B.S. Code of Practice 114—1957 which gives both volume and weight.

Normal Mix (by volume)	*Cubic feet of aggregate per 112 lb of cement*	
	Fine	*Coarse*
1:1:2	1¼	2½
1:1½:3	1⅞	3¾
1:2:4	2½	5

The choice of a mix will depend upon the class of work for which it is to be used, e.g. the quality of the concrete in foundation work will not require to be as good as in a reinforced concrete structure. In certain classes of work an ungraded mix is sometimes required, in which case '*all in*' *ballast* is used, which will consist of fine and coarse aggregate excavated from the ground in its natural proportions, to which cement is added. This type of concrete is inferior to the graded concretes.

Water for Concrete Work

Clean water must be used and the strength of concrete will depend upon the ratio of water to cement in the mix. The excessive addition of water will weaken the concrete and the builder should be aware of this, particularly in relation to workability, i.e. the wetness of the mix conducive to even spreading. It is important, therefore, in any construction work which relies on the strength of the concrete, that the water added to the mix is controlled. For smaller foundation work the water added is not so critical, but, even so, should not be excessive. It should give sufficient workability to place the concrete in position. It is outside the scope of the subject at this stage to consider the effect of additives and the vibration of concrete as related to water content, but this will be considered at later stages of the subject.

Mixing

Most mixing of concrete will be by machine. Concrete mixers range from the small types with a capacity of 3½ T (tilting type) to large batching plants, comprising large mixers using mechanical loading of aggregate and silo fed cement. The builder will usually have a mixer, suitable for the class of work which he undertakes, for transporting to the site. Mixers are of the tilting and non-tilting type. The coarse and fine aggregate will be placed in the hopper drum in defined proportions, the cement and water are added and the whole mixed by rotation of the drum.

Should it be required to hand-mix the concrete the mixing area must be clean and the dry materials turned over twice, the water added and turned over twice more.

Concrete when mixed will have two setting times, the first, called the *initial setting time*, should not be less than 30 minutes after mixing for ordinary Portland Cement, the second called the *final setting time* should be not more than 10 hours after mixing.

Placing

The mixed concrete will be conveyed from the mixer to the position it is to occupy by barrow or dumper. It should not be placed on the ground but directly into position, before the initial setting has taken place, otherwise the concrete will suffer damage. The concrete will be tamped, i.e. compacted by rodding to remove all air pockets, etc., and finished level on top with a spade or wood float.

Concrete Strip Foundation

The purpose of the foundation is to spread the load from the wall above, over an area to suit the bearing capacity of the subsoil. The width of the foundations must not be less than the width of the wall and will be up to 3′ 0″ for very poor subsoil. Where a width is required greater than this, special methods will have to be used.

The Building Regulations give the widths of foundations for various classes of subsoil.

The depth of concrete in the foundation will be 6″.

Oversite Concrete

It is a statutory requirement that the area between the external walls of habitable buildings is covered with concrete or other suitable material. It is usual to cover the area with a minimum of 4″ of concrete.

The purpose of this covering is to prevent damp rising within the building which will cause unhealthy conditions and have a detrimental effect on timbers used in the ground floor construction. See Figure 14.

3 Walls

THE function of a wall in a domestic building is:

1. To provide a compartment for living or working.
2. To resist weather, particularly damp and cold.
3. To resist the passage of noise.
4. To safely carry the structural loads placed upon it.
5. To be reasonable in cost.

There are several materials which can be used in the construction of walls, but for domestic work bricks are still the most popular choice and provide the most economical type of structure. Other materials provide their own particular function in relation to buildings, such as offices, factories and schools, etc.

Bricks

Bricks can be manufactured from clay, lime and sand and various other aggregates, the process being different between clay and the other materials.

Clay bricks are manufactured from clay and sand in defined proportions, the type of brick varying according to the proportion of their ingredients, e.g. the greater the clay content the more dense the brick, the greater the sand content the softer the brick, making it easier to cut, mould or rub.

Bricks in the past have been manufactured in the localities where clay was found and these small manufacturing units supplied the bricks for the locality and many kilns, now disused, can be seen in various parts of the country. With the development of transport larger manufacturing units were established, supplying bricks to any part of the country at economic prices, thus causing the smaller units to close or to specialise in facing bricks of a special quality and appearance.

The small brickyards use a kiln of the older type for burning the bricks, but the larger yards have adopted the *Hoffman kiln* as being the most economical. Clay is excavated from a pit by a dragline, or mechanical navvy, the different types of clay being separated for use in common bricks or facing bricks. The excavated shale is placed into continuously moving trucks and transported to the grinding mill. The ground material is raised on to revolving screens, which deposit the screened clay on to a steel band to convey it to either a hopper of the brick machine, or to an extruding plant, which is fed by a screw in a similar manner to a mincing machine. The extruded clay strip is cut into bricks by wires. These can be made into pressed bricks by placing in a press, which will also provide the '*frog*' (see Figure 5(a)).

The Hoffman kiln is built in the form of U on plan and divided into compartments or chambers, as shown in Figure 5(d). The process of the kiln is continuous, the newly made bricks being stacked on each other with an inch space between each brick. Hot air from the kiln passes over the green bricks to dry them and the chamber is slowly heated raising the temperature to fire the bricks and then allowing them to cool. Each chamber in the kiln has this rotation of heat so that the bricks are not moved but the firing rotates around the kiln from chamber to chamber. After firing and cooling the bricks are removed, the process taking about seven days to fire to a temperature of 1500°F and a further six days to cool. This process means that one or more chambers can be loaded and drawn each day.

Hand-made bricks are formed in timber moulds, removed, dried and burnt, the blemishes on the face being a feature of the brick.

Choice of Bricks and Blocks

It is important that bricks should be chosen for the position in which they are to be placed. This will relate to their colour, strength and damp resistance. A common brick for domestic work does not require to be too strong. Should the wall require to carry a heavy concentrated load, a stronger brick may be necessary to resist the crushing effect. Facing bricks will be chosen mainly for their appearance while in basement work the bricks may be chosen for their resistance to damp penetration.

Blocks will be used as partitions or as internal skins to hollow walls where appearance is usually secondary. They will need to be strong and, if necessary, sound and heat resisting to prevent movement of sound and heat from one room to another.

Sand-Lime Bricks

Sand-lime bricks are manufactured from a mixture of sand and lime in defined proportions, placed in a machine and moulded to shape, stacked on a movable truck and passed into a steam chamber to cure. A white pleasing brick is formed.

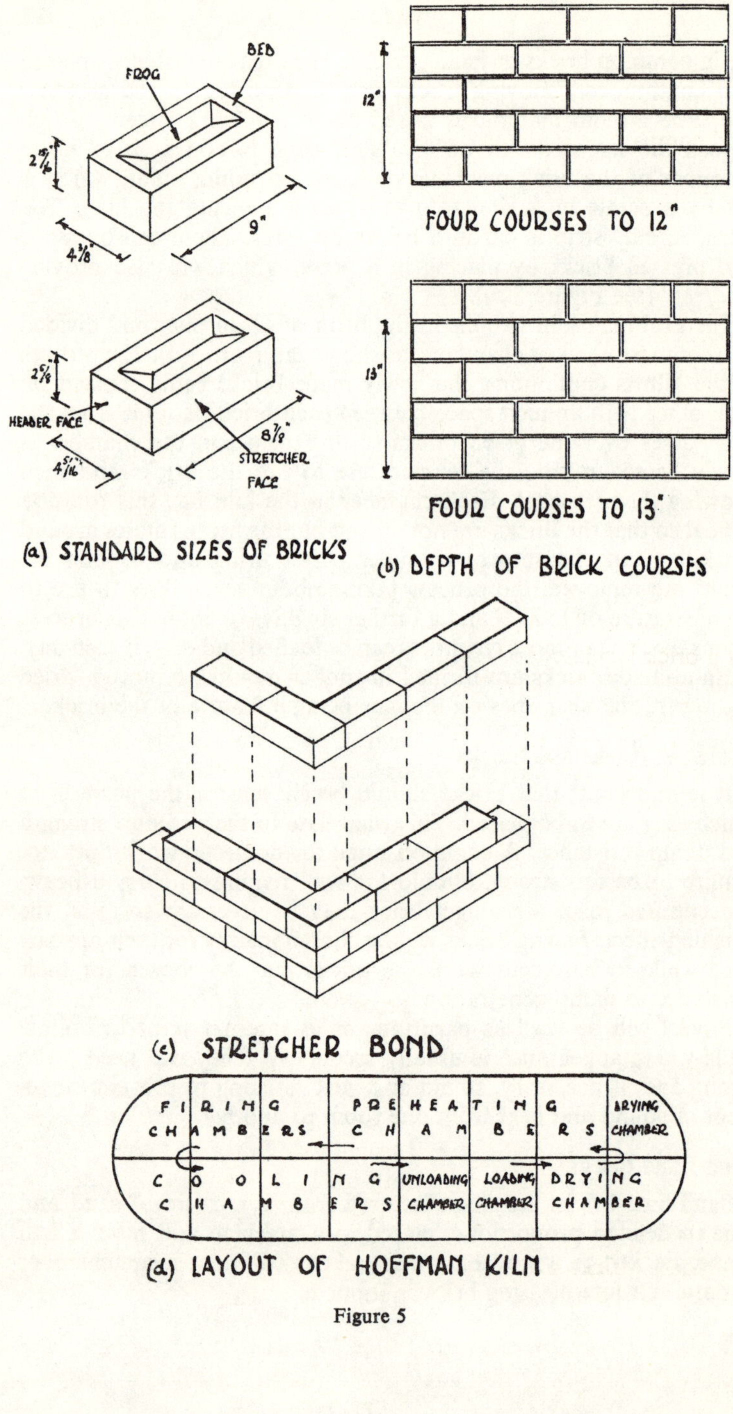
BED
FROG
2 15/16"
9"
4 3/8"
2 5/8"
HEADER FACE
8 7/8"
STRETCHER FACE
4 5/16"
(a) STANDARD SIZES OF BRICKS
12"
FOUR COURSES TO 12"
13"
FOUR COURSES TO 13"
(b) DEPTH OF BRICK COURSES
(c) STRETCHER BOND
FIRING CHAMBERS
PREHEATING CHAMBERS
DRYING CHAMBER
COOLING CHAMBERS
UNLOADING CHAMBER
LOADING CHAMBER
DRYING CHAMBER
(d) LAYOUT OF HOFFMAN KILN

Figure 5

Concrete Bricks

Bricks manufactured from clinker have been used successfully and compare very favourably with ordinary clay bricks. No ill effect will be caused by using in conjunction with other material provided the aggregate is carefully selected, and the brick tested for moisture movement.

Classification of Bricks

Bricks can be divided into various classes and uses according to their properties.

Common bricks are used for general building work when appearance is not vital.

Facing bricks are used for appearance, there being many varieties.

Engineering bricks are used for positions in buildings where strength is required or damp penetration is to be resisted.

Brick Sizes

When bricks were manufactured in the various small units over the country the size of the brick varied according to the firm making them, but even so this variation was within limits, because the size of a brick is related to the ability of the craftsman to be able to raise and place the brick in position with one hand, a large size brick would require two hands to raise it, the trowel having to be laid down to do so. The R.I.B.A. endeavoured to standardise the size of bricks and this standard has been incorporated in the British Standards and manufacturers are requested to conform when building new plant or renewing existing plant. The general difference in size between the north of the country, where 3″ and $2\frac{7}{8}$″ deep bricks are used, and the south where the $2\frac{5}{8}$″ deep brick is used, has been adhered to in the standard size brick.

The standard sizes of bricks are given in BS. 657—1962 and are illustrated in Figure 5(a) and (b), which also shows the course heights.

Bricks will vary in weight according to their composition, the greater the clay content the heavier the brick.

Some bricks have indentations in their bed known as '*frogs*' as shown in Figures 5(a) and 8(d). This will make the brick lighter, easier to handle and will also provide a key for the mortar. The frog shape will vary with the make, but is usually of V cross-section, and may be in either the top bed or in both top and bottom beds.

Special Bricks

Where bricks are required for special purposes, purpose-made bricks will be used, e.g. squint quoins, plinths, etc.

Keyed Bricks

Where bricks are to be plastered after building a keyed brick may be used. Vertical grooves are formed in the face and head of the brick to provide for the key, as shown in Figure 8(c).

Arch Bricks

Bricks to be built into arches in faced work will have to be wedge shaped to allow for the curve. Special bricks are manufactured for arches, or bricks with a large sand content can be rubbed to shape, these are known as 'rubber' bricks.

Engineering Bricks

Bricks having a large clay content are dense, making them suitable for positions when damp has to be resisted or heavy loads are to be supported on the walls.

Mortars

The choice of mortars for brickwork must also be considered. Generally speaking, a mortar should not be stronger than the brick being used. Where mortars are stronger any slight settlement which might take place will cause the brick to crack, while when the mortar is weaker than the brick, the joint will crack instead and will be easier to fill than the crack in the brick.

There are three main types of mortar:

(*a*) *Lime mortar* is composed of hydrated lime and sand and provides a weaker mortar, being suitable for use with the softer brick.

(*b*) *Cement mortar* is composed of Portland cement and sand and is commonly used today. It is stronger than lime mortar and is particularly suitable for work where damp is to be resisted.

(*c*) *Gauged mortar* is composed of Portland cement, hydrated lime and sand and gives a mortar of strength between the two former types and is easier to spread than cement mortar.

The sand, or fine aggregate, must be clean and washed, free from clay or other foreign matter.

Building sand is commonly used, which is a soft sand, i.e. it is not so angular and gritty as concreting sand, making it easier to spread.

Coloured cement can be used instead of ordinary Portland cement, the use of which will alter the appearance of the facings, making

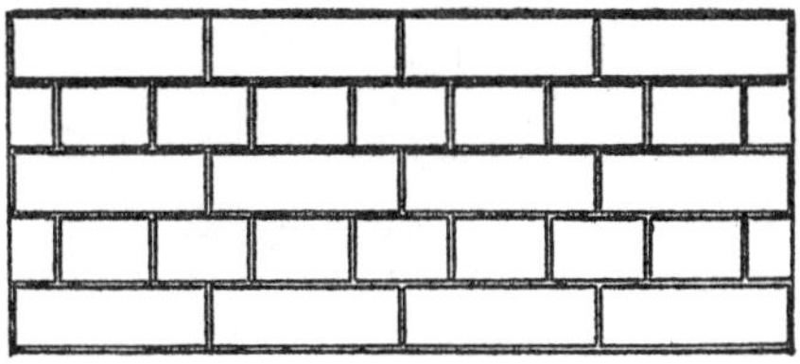

(a) ELEVATION OF ENGLISH BOND

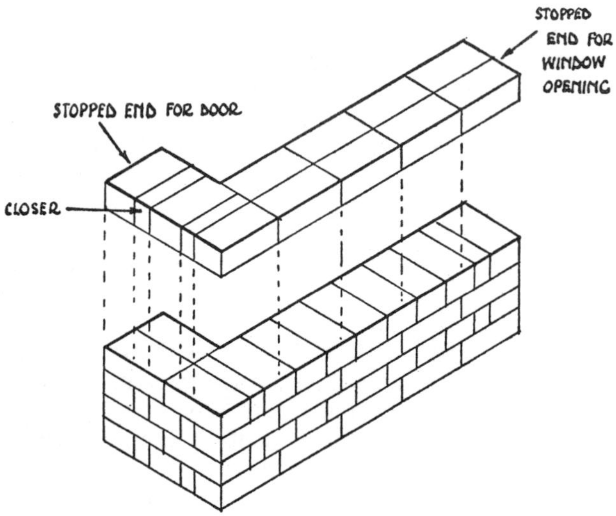

(b) SKETCH OF ONE BRICK WALL BUILT IN ENGLISH BOND

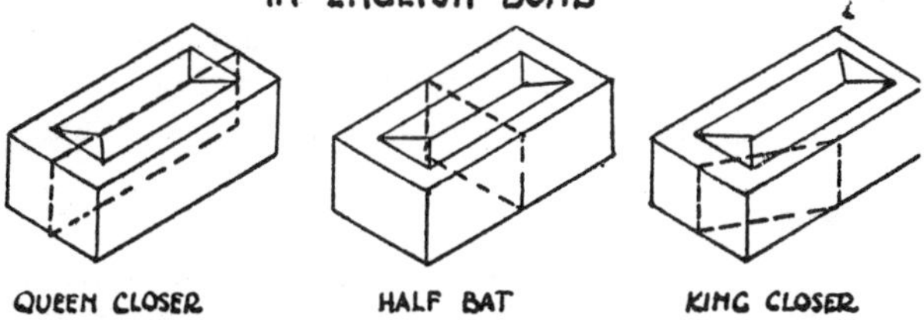

(c) CUTS ON BRICKS FOR BONDING

Figure 6

the brickwork become interesting or otherwise, by emphasis on the joint or by toning in with the brick colouring, and care must be taken to ensure that the correct effect is obtained.

Common mixes for mortars are as follows:

Lime Mortar	1:3 or 1:4
Cement Mortar	1:3 or 1:4
Gauged Mortar	1:3:12 or 1:1:6

Mixing of Mortar

With the introduction of the smaller mixer, mortars are usually mixed by machine in a similar manner to that described for concrete.

Bonding

The purpose of bonding in the older building was to give strength to the wall by the interlacing of the bricks, in order to spread the load which the wall had to support, with the elimination of continuous vertical joints, which gave a weakness. With the introduction of Portland cement the need for bonding for strength has been eliminated, as the mortar is often as strong as the brick. The use of bonding in the past for strength has developed a face pattern and bonding today is used by the architect to give a desired face appearance.

The main bonds in use are:

Stretcher Bond, used in half brick walls, as shown in Figure 5(c).

English Bond. The strongest bond of the three given, is used for walls over $4\frac{1}{2}''$ thick, as shown in Figure 6(b).

Flemish Bond may be 'single' or 'double'. The more common is double Flemish, as shown in Figure 9(a).

Single Flemish is only suitable for thicker walls.

The theoretical weakness of Flemish bond is the $2\frac{1}{4}''$ vertical joint as indicated in Figure 9(a). The unit form of the bond shows the basis of the bond.

In English and Flemish bond the provision of *a closer* adjacent to the corner header will provide the break of bond between each course. Figure 6(c) shows the cuts on bricks for bonding.

Cavity Walls

These have become the normal form of construction of walls in housing work. The walls consist of two skins or leaves of half brick thickness with a 2″ cavity between, as shown in Figure 7(a). To strengthen the two skins and make the wall act as a unit, *wall ties* are built into the two skins. Care must be taken in the choice and

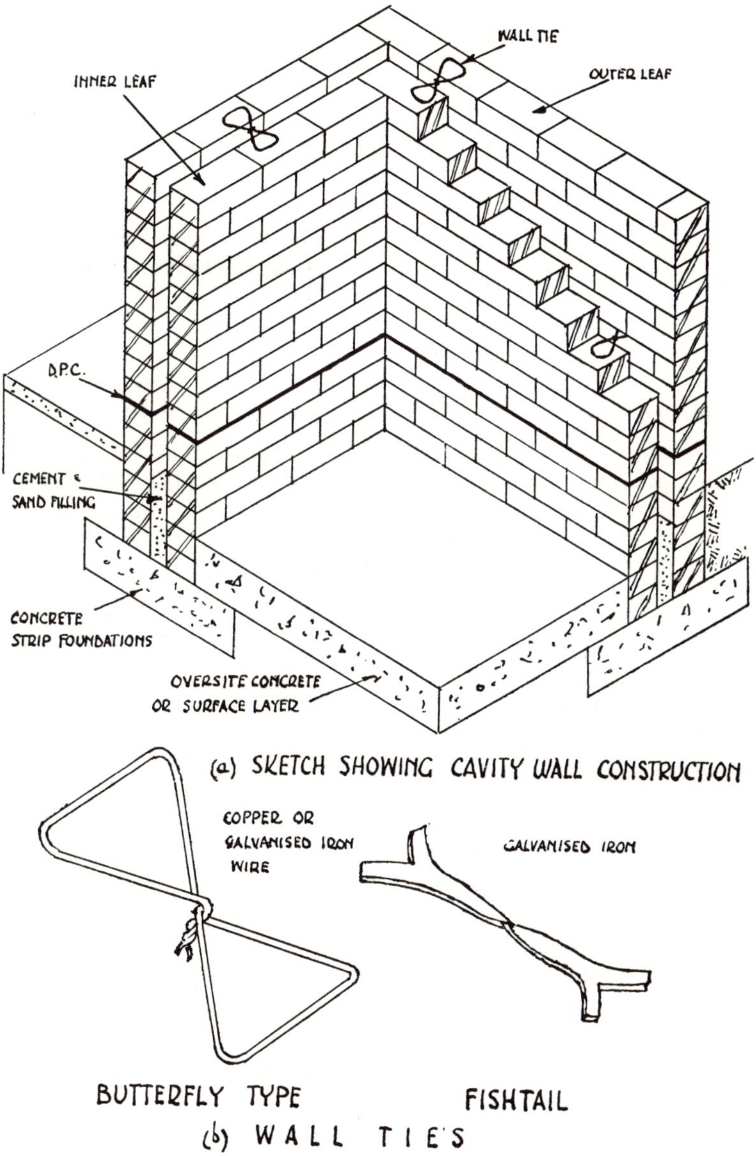
WALL TIE
INNER LEAF
OUTER LEAF
D.P.C.
CEMENT & SAND FILLING
CONCRETE STRIP FOUNDATIONS
OVERSITE CONCRETE OR SURFACE LAYER
(a) SKETCH SHOWING CAVITY WALL CONSTRUCTION
COPPER OR GALVANISED IRON WIRE
GALVANISED IRON
BUTTERFLY TYPE
FISHTAIL
(b) WALL TIES

Figure 7

building in of wall ties. The outer skin of the wall will become saturated during wet periods of the year, and to prevent the passage of moisture to the inner skin the wall tie must be designed correctly. Suitable types of wall tie which conform to the BSS requirements are given in Figure 7(b). Wall ties should be staggered in the wall so that any water dropping from the tie will miss the tie in the row below.

A 9″ thick solid wall may allow moisture to pass to the inner face, particularly if the facing bricks are soft and porous. The cavity wall of the same thickness will have a two-inch cavity barrier to damp.

The cavity can also be used as a barrier to sound and thermal movement, particularly if a lightweight or thermal block is used in the inner skin and the cavity is sealed.

Pointing

Many of the older forms of pointing have given way to the modern flush pointing, which is cheaper and more in keeping with trends in architecture. Figure 8(e) shows several types of pointing and the correct manner in which they should be carried out. Care must always be taken to prevent water collecting in joints, otherwise, when the water freezes it may break off the brick at that point and cause the whole appearance of the wall to be spoilt.

Efflorescence

This is the white powder deposit sometimes found on dry brick walls, which will disappear after a shower of rain, only to re-appear after drying again.

This may be caused by one of three things:

1. Salts in the brick, due to salts in the clay or soil prior to manufacture. The manufacturers normally guard against this.
2. Salts in the mortar due to salts in the materials, i.e. sea sand or water. Care should be taken in the selection of the source of the sand.
3. Salts in the ground where the building is being erected. An adequate damp-proof course should prevent the passage of moisture from the ground to the brickwork above, but the work below the damp-proof course may be subject to efflorescence.

When efflorescence has entered the brickwork it is very difficult to eradicate even with constant washing, and it should be guarded against at its source.

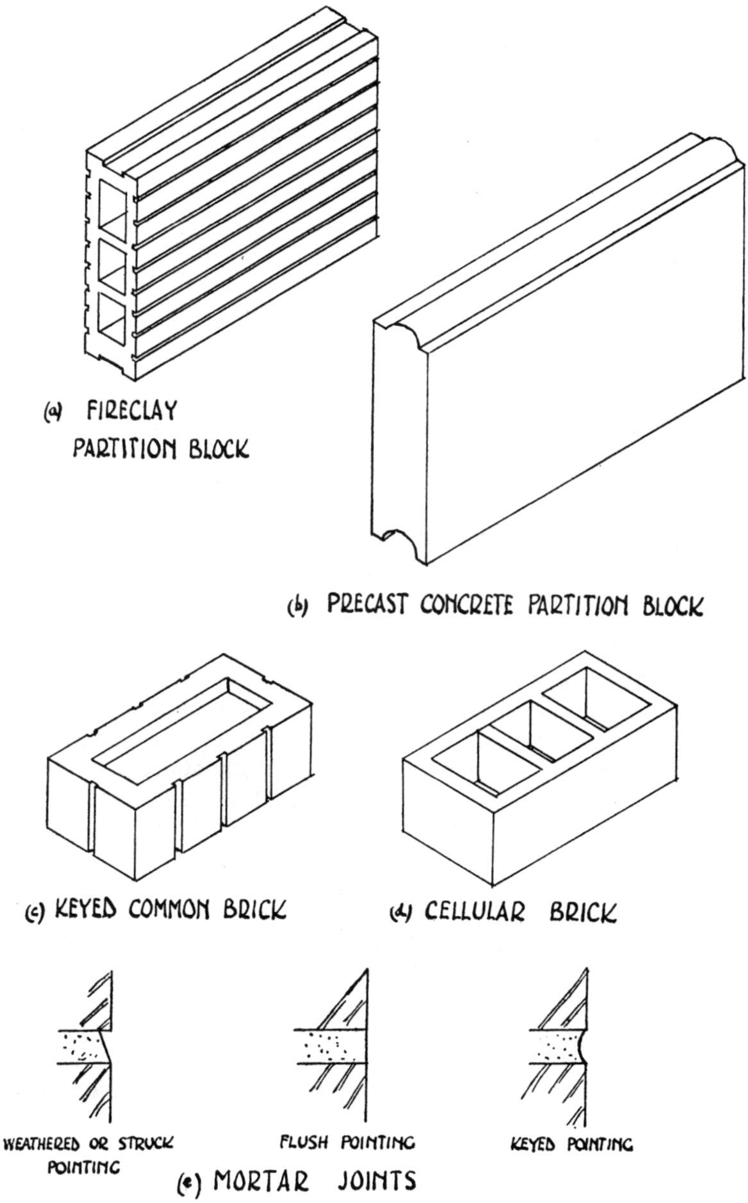
(a) FIRECLAY PARTITION BLOCK
(b) PRECAST CONCRETE PARTITION BLOCK
(c) KEYED COMMON BRICK
(d) CELLULAR BRICK
WEATHERED OR STRUCK POINTING
FLUSH POINTING
KEYED POINTING
(e) MORTAR JOINTS

Figure 8

Blocks

There has been an increased use of blocks in domestic work for the inner leaf of cavity walls and for internal partitions. Blocks have certain advantages over bricks because of their larger size and reduction in cost. There are several types of blocks in use, the following being the most common:

Concrete Blocks

These are composed of coarse and fine aggregate and cement in a controlled mix under factory conditions. The mixed material is contained in moulds and pressure, or vibration, is used to consolidate, enabling a dry mix to be used. The bed of the block can be keyed to strengthen the joint. See Figure 8(b).

Terracotta Blocks

These blocks are composed of fireclay and are manufactured in a similar manner to bricks. The blocks are made lighter by being formed with a hollow centre as shown in Figure 8(a) and the outer surface is keyed by indentations in the block.

Special filler pieces are provided for the ends of blocks which are exposed at breaks in the wall.

Lightweight Concrete Blocks

Special proprietary made blocks are offered on the market. The blocks will be composed of a lightweight aggregate with cement and will set in a similar manner to concrete. The common aggregate is foamed slag and furnace residue. As already stated for concrete bricks, care must be taken in the choice of the aggregate to ensure that no chemical action will take place when plaster or other finishes are applied to it.

Scaffolding

Only a light scaffolding will be required for a one-storey building. For the greater proportion of the walling and internal partitions a simple trestle scaffold will be sufficient, but for the gable walling and chimney stack a light tubular scaffold will be necessary. It should be noted that the trestle scaffold must be stable. The supporting of scaffold boards on drain pipes or similar unstable supports should not be used.

Parapet Walls and Copings

Where flat roofs are constructed it is common for the main walls to be continued above the flat roof to form a parapet wall. The

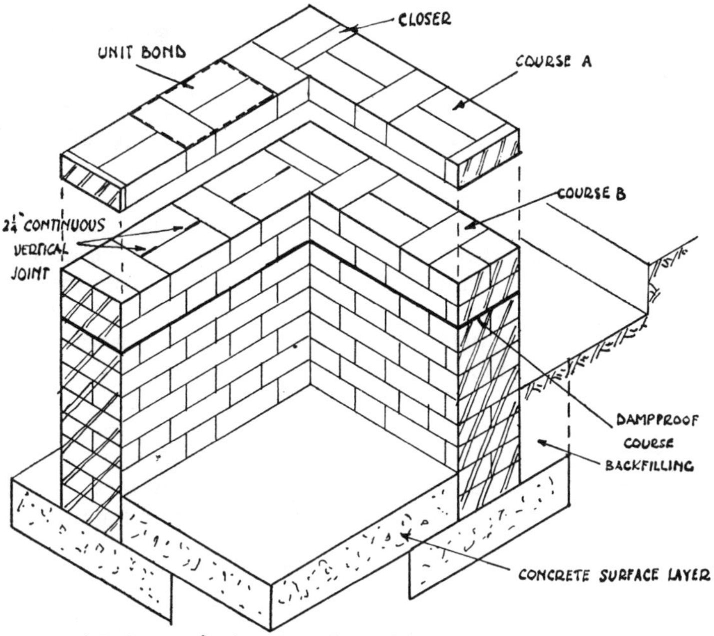

(a) SKETCH OF 9" WALL BUILT IN FLEMISH BOND

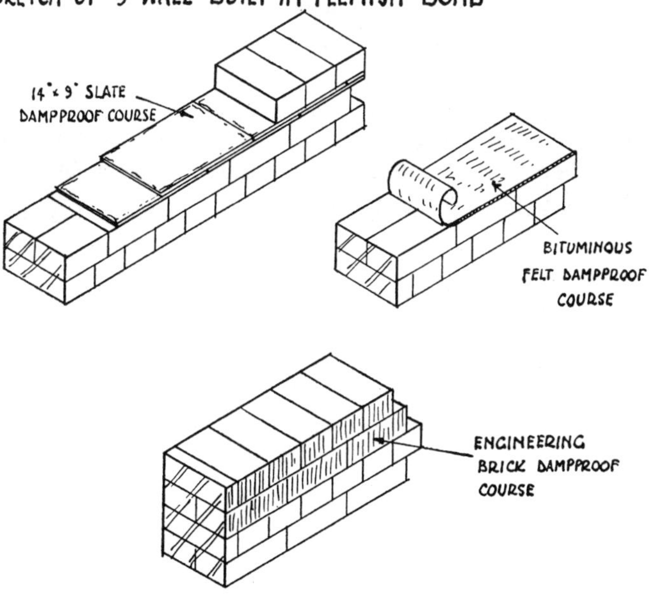

(b) COMMON TYPES OF DAMPPROOF COURSES

Figure 9

thickness of the wall will be 9″ solid brickwork, the top being finished with a coping. The purpose of this coping is to prevent excessive saturation of the top of the wall. A damp-proof course should also be provided through the wall to prevent downward movement of the moisture in the wall.

Copings may be of brick, or natural or reconstructed stone. Brick copings will be flush with the wall on both sides, but the addition of a tile creasing will throw the surplus water away from the wall.

Stone copings should be designed to overhang and be weathered to direct the surplus water towards the roof surface, where it will be collected in the rainwater gutter. The underside of the coping should be provided with a throat on the inner side of the wall, to prevent the water running back on to the wall. Figures 26(d) and (e) show a parapet wall with simple brick and stone copings.

Copings may also be used on boundary walls, but these will not require weathering and the throat will be provided on both sides of the wall.

Damp-proof Courses

Every wall must be provided with an efficient damp-proof course of impervious material. It should be 6″ above ground level and may consist of one of the following materials:

(*a*) Two courses of engineering brick.
(*b*) Two courses of stout slate in cement mortar.
(*c*) Bituminous felt or one of the proprietary felts with lead, copper or aluminium foil interleaved with the felt.

Sketches of the damp-proof courses are shown in Figure 9(b).

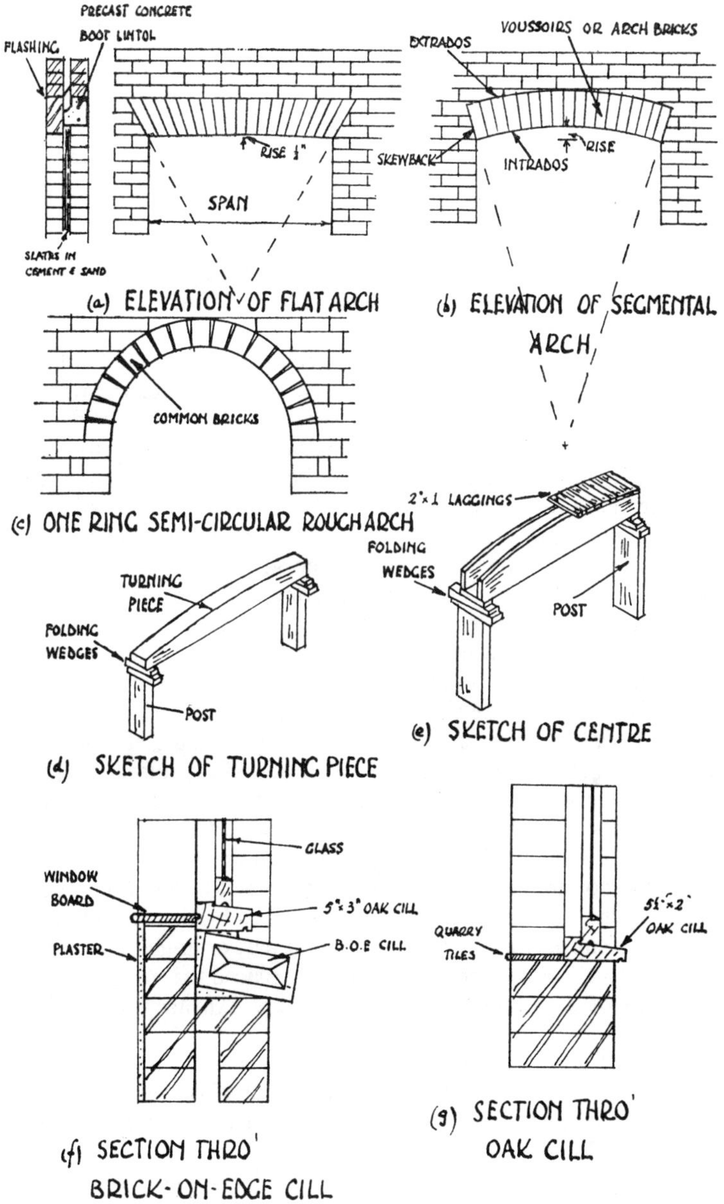
PRECAST CONCRETE
BOOT LINTOL
FLASHING
SLATES IN
CEMENT & SAND
RISE ½"
SPAN
(a) ELEVATION OF FLAT ARCH
VOUSSOIRS OR ARCH BRICKS
EXTRADOS
SKEWBACK
INTRADOS
RISE
(b) ELEVATION OF SEGMENTAL ARCH
COMMON BRICKS
(c) ONE RING SEMI-CIRCULAR ROUGH ARCH
2"×1 LAGGINGS
FOLDING
WEDGES
POST
TURNING
PIECE
FOLDING
WEDGES
POST
(e) SKETCH OF CENTRE
(d) SKETCH OF TURNING PIECE
GLASS
WINDOW
BOARD
5"×3" OAK CILL
PLASTER
B.O.E CILL
QUARRY
TILES
OAK CILL
(g) SECTION THRO' OAK CILL
(f) SECTION THRO' BRICK-ON-EDGE CILL

Figure 10

4 Openings in Walls

It must be remembered that any opening causes a potential weakness in a wall, not only in its strength but also in its weather resistance and this must be considered when studying openings. The work above the opening must be supported. In ascertaining the load to be carried a triangulated area of 60° may be calculated for the width of the opening.

LINTELS. A lintel is a common method of supporting the brickwork above the opening. Lintels may be of timber or of concrete, the latter being the more usual and may be *in situ* or precast. *In situ* concrete lintels are cast in position, while precast concrete lintels are either made in a factory, or on the ground on the site, being hoisted and fixed over the opening. Concrete lintels must be reinforced with mild steel reinforcing rods as shown in Figure 11(b). The rods will be placed in the lintel at the bottom, to resist tension, with sufficient cover of concrete on the underside to protect the steel.

ARCHES may also be used to support the brickwork above. They can be of several types, flat, segmental, semicircular, elliptical, etc.

A flat arch is shown in Figure 10(a). The soffit is raised slightly to overcome the appearance of sagging. The bricks are tapered to support each other, the line between the brickwork and the arch being termed the *skewback*. The bricks of the arch are termed *voussoirs*.

Another common form of the flat arch is the '*brick-on-end*' or 'soldier' arch. This arch does not have tapered bricks and, therefore, will require supporting by placing a metal bar, or angle, underneath, or, alternatively, a reinforcing rod may be placed through the bricks. The depth of a 'soldier' arch is 9″ on face. An arch and a lintel behind are shown in Figure 11(a).

Segmental arches have a curved soffit as shown in Figure 10(b). The depth of segmental arches may be 4½″ or 9″ on face.

Semicircular and other curved arches will follow the pattern of the segmental arch.

Rough arch. When an arch is provided for support only and is being covered with plaster or other facing material, it can be formed of common bricks in one or two rings as shown in Figure 10(c).

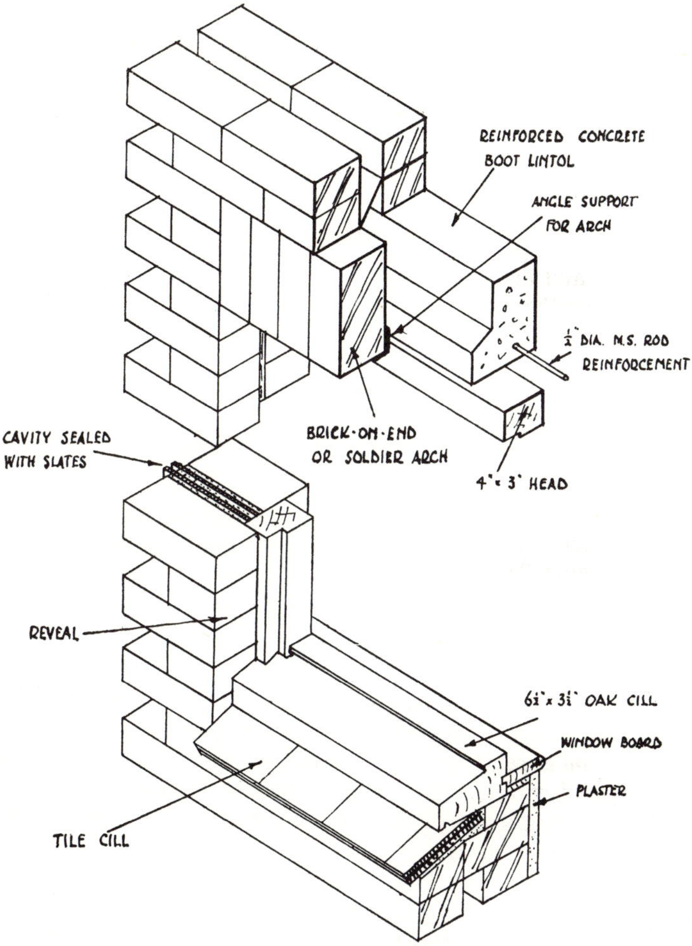

(a) DETAIL OF WINDOW OPENING

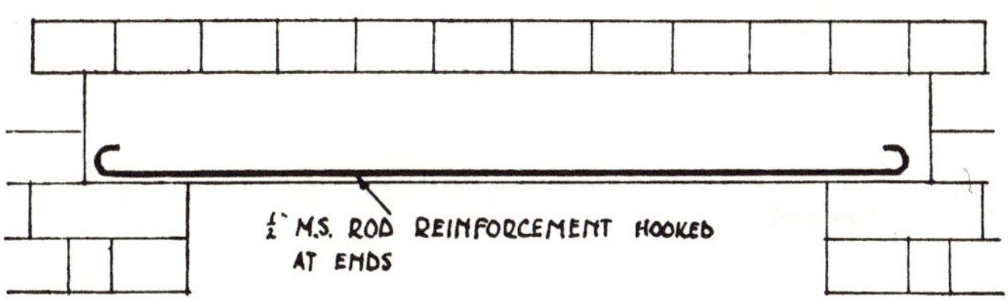

(b) LINTOL OVER OPENING SHOWING REINFORCEMENT

Figure 11

Fair arches. Where special arch bricks, or facing bricks, are used in arches, these are termed fair arches.

Figures 10(d) and (e) show temporary supports for arches during construction, for $4\frac{1}{2}''$ wide arches termed a 'turning piece' and wider arches a 'centre'.

Reveals

The sides of an opening are termed 'reveals' and must be formed with a vertical face, internally they may be covered with plaster, but externally will usually be in faced bricks. To provide for this the brickwork should be bonded in a similar manner already given for bonding to quoins and stopped ends.

Suitable bonding to reveals is given in Figures 28, 30 and 31. Reveals should be rebated to receive the door or window frame, this will protect the opening against weather penetration.

Sills

The lower part of the opening must be finished to receive the window and also throw the water, which collects at this position, away from the opening. It should be remembered that the large area of glass in windows, which will be impervious to moisture, will require the sill to take a considerable quantity of water. There are several methods of dealing with this.

(*a*) *Stone sill.* This may be in natural or reconstructed stone but will, in many cases, be too expensive for normal domestic work. A typical section through a stone sill is shown in Figure 28. Stone sills should be bedded hollow so that they do not crack when the brickwork is built to the jambs of the opening. The underside of the sill will be pointed afterwards.

(*b*) *Tile sill.* Can form a useful sill for both timber and metal windows, as shown in Figure 11(a).

(*c*) *Timber sill.* The use of the timber sill in conjunction with the window is often used but will reduce the external reveal. See Figure 10(g).

In each case the top of the sill must be weathered, or sloped, towards the outer face of the building. It should also project beyond the face of the wall so that the water is thrown clear. The underside of the projecting portion of the sill should have a throat, or drip, to prevent the water running back to the wall.

When two materials form a sill and sub-sill, such as stone and timber, a water bar should be provided to prevent movement of water through the wall. The water bar will be galvanised iron $1'' \times \frac{1}{4}''$ bedded in red lead, as shown in Figure 28.

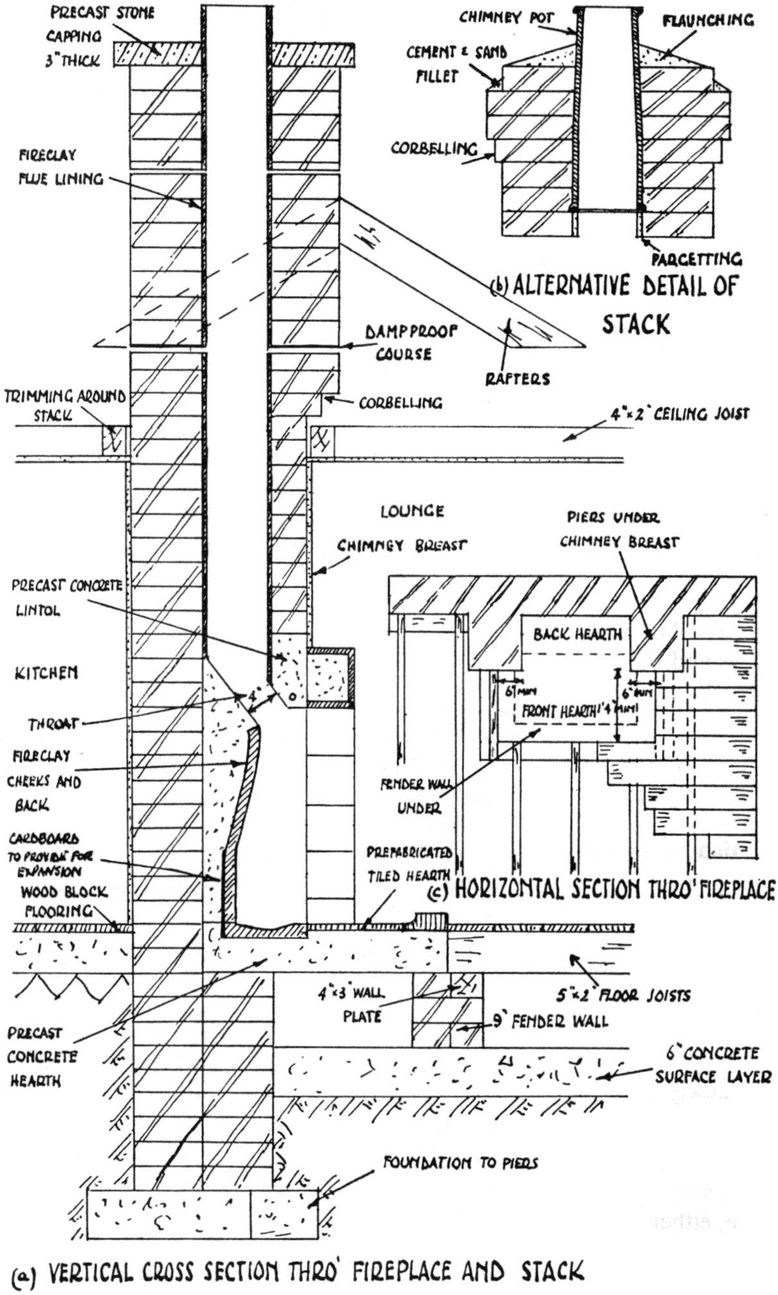
PRECAST STONE CAPPING 3" THICK
FIRECLAY FLUE LINING
DAMP PROOF COURSE
RAFTERS
CORBELLING
TRIMMING AROUND STACK
4"×2" CEILING JOIST
LOUNGE
CHIMNEY BREAST
PRECAST CONCRETE LINTOL
KITCHEN
4"
THROAT
FIRECLAY CHEEKS AND BACK
CARDBOARD TO PROVIDE FOR EXPANSION
WOOD BLOCK FLOORING
PREFABRICATED TILED HEARTH
PRECAST CONCRETE HEARTH
4"×3" WALL PLATE
5"×2" FLOOR JOISTS
9" FENDER WALL
6" CONCRETE SURFACE LAYER
FOUNDATION TO PIERS
(a) VERTICAL CROSS SECTION THRO' FIREPLACE AND STACK
CHIMNEY POT
FLAUNCHING
CEMENT & SAND FILLET
CORBELLING
PARGETTING
(b) ALTERNATIVE DETAIL OF STACK
PIERS UNDER CHIMNEY BREAST
BACK HEARTH
6" MIN
FRONT HEARTH 1'4" MIN
FENDER WALL UNDER
(c) HORIZONTAL SECTION THRO' FIREPLACE

Figure 12

5 Chimney Construction

REFERENCE should be made to the Regulations in construction of chimneys and these will be included in the following outline of their construction.

Situation of Chimneys

This will be determined in the design of the building, but when on an external wall may bring the stack at the eaves, while if on an internal wall, will bring the stack through the roof surface or ridge.

Foundation

The construction of foundations to walls also applies to chimneys. When the chimney is part of an external or internal wall, the construction of the chimney from the foundation to the top of the ground floor fireplace opening will be two attached piers, and the foundation to the main wall will project to form the foundation to the piers, as shown in Figure 12(a).

Ground Floor Fireplace

Few newly constructed domestic buildings have basements, so that the surface layer in the case of timber joisted ground floors can be used to support a *fender wall* around the fireplace. The purpose of a fender wall is to provide support for the hearth and the floor joists, if necessary.

The hearth, which must be of incombustible material, will be of concrete, either precast or *in situ*, 6″ thick, the top surface not being lower than the floor level, it should project from the breast at least 16″ and extend 6″ on either side of the opening. A suitable tile, or other finish, will be fixed on the top surface, or a precast tiled hearth will be used in conjunction with the fireplace surround.

The width and height of the fireplace opening will depend upon the size of the room.

The projecting brickwork above the ground floor fireplace opening will be termed '*the breast*' and will be supported on a concrete lintel or rough arch. The lintel should be shaped as shown on Figure 12(a) and the back of the opening brought forward to restrict the flue at this point. The width of the throat should not be greater than 4″, the length being related in area to that of the flue, i.e. for 9″ × 9″ flue, which is usual in domestic work, the length of the throat will be 21″, giving approximately the same area of opening.

In solid floor construction the same treatment of the hearth and fireplace can take place.

The flue from the fireplace must be either pargetted or lined. Pargetting is the application of cement and sand rendering to the inside of the flue as the chimney is built. To prevent droppings of mortar collecting at the base of the flue a sack of straw may be raised as the chimney is built and while pargetting is applied. A more satisfactory lining of the flue is the use of fireclay flue liners which can be either square or round. These are built into the flue when the chimney is constructed.

In both cases the purpose of the finish to the flue is to provide a smooth surface so that soot does not collect, and also for easier sweeping. In the case of slow burning fires or boilers the use of a flue liner is necessary to prevent the flue from cooling too quickly and the air within becoming sluggish, which will allow the gases in the smoke and fumes to attack the brickwork.

In the design of flues the smoke should discharge into the open air as quickly as possible and bends in the flue should be eliminated as far as practicable. This will reduce the possibility of back draught and smoky fires. Each fireplace must have its own flue and no flue may discharge into another, but must be taken to the open air.

Chimney Stack

The chimney breasts passing through the roof space will be reduced to form the stack, the size of which will be determined by the number of flues to be accommodated. While it is better to have 9″ brick surround to the stack with 4½″ divisions between flues, known as 'withes' or 'midfeathers', with the use of flue liners 4½″ thick surround is sufficient for domestic work.

The termination of a stack may be of brickwork or with a precast concrete or natural stone capping. The flue will terminate in a chimney pot, of which there is a great variety. The height of the pot which is to be visible will depend upon the design of the building. With pargetted flues the pot must be supported and one method is shown in Figure 12(b). Where flue liners are used the pot will rest upon them.

The top of the stack should be weathered and if cement and sand are used for this purpose, as shown in Figure 12(b), this is known as '*flaunching*'.

A damp-proof course should be included in the stack within the depth of the roof space, as indicated in Figure 12(a). Chimney stacks are one of the wettest parts of the building because, from whichever direction the rain drives, the stack will be exposed. The damp-proof course will prevent the water penetrating the interior of the building thereby causing damage to structure and finishings.

6 Timber

For domestic purposes softwoods will be used for this work, except for sills or where a superior class of work is required, such as in Banks, Insurance Buildings, etc., in which case a hardwood will be used. The choice will be made according to the amount of money available and the need for strength or durability.

Timber grows outwardly adding further layers of growth each year, the type of growth varying with the environment of the tree during growth. This layer of growth which takes place is often termed '*an annular ring*'. A sketch through a tree is given in Figure 13(g) and the various terms are given.

During the spring and autumn the tree will have moisture movement within it and the formation of the *cambium layer* takes place. The spring wood is immature, but the autumn wood has matured during the summer period.

Seasoning

Since the growth of the tree is related to the movement of sap, before a felled tree can be used it must have the moisture within it reduced. The amount of moisture reduction will be related to the use of the timber in the construction, since further movement may take place according to the amount of heat or dampness to which the timber will be subjected, after being fixed in the structure.

Timber should be felled in the winter period when the sap is at its lowest. It should be cut into logs or boards, stacked and protected from the sun and rain but with adequate ventilation. This is termed natural seasoning, which is slow.

Kiln seasoning is much quicker and more economical and consists of stacking the timber on trolleys and passing into a kiln for scientifically controlled drying. The drying should not be too quick and should consist of a circulation of warm air, which will be controlled for humidity and will be continuously changed. The humidity will be gradually reduced until the timber has the required moisture content.

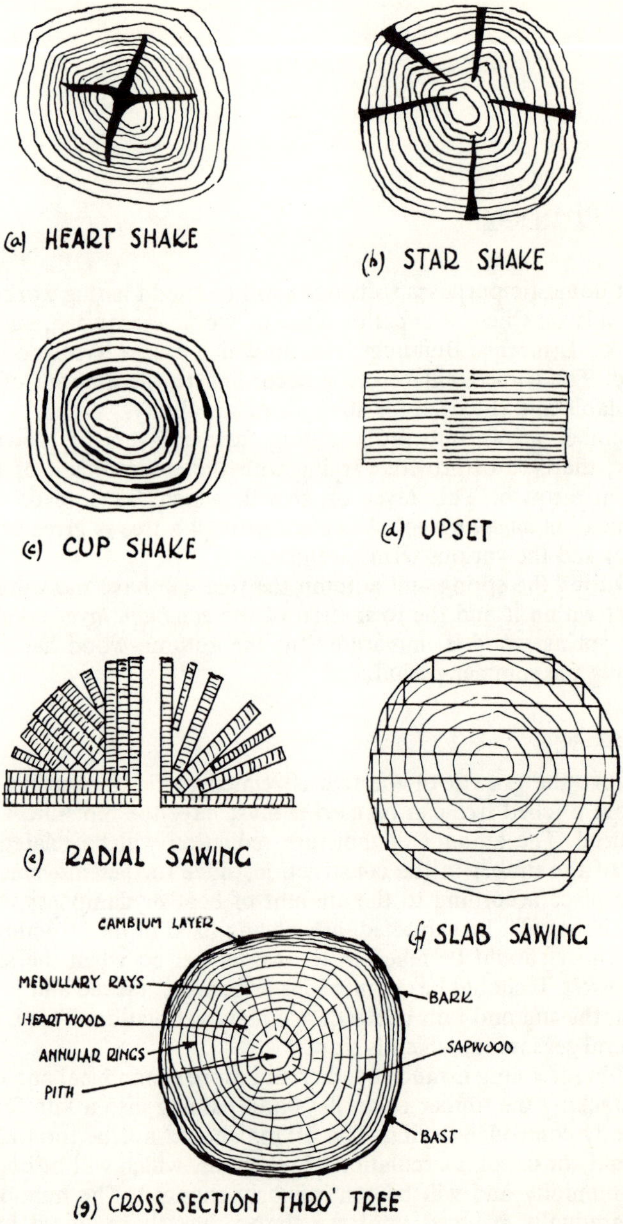
(a) HEART SHAKE
(b) STAR SHAKE
(c) CUP SHAKE
(d) UPSET
(e) RADIAL SAWING
(f) SLAB SAWING
CAMBIUM LAYER
MEDULLARY RAYS
HEARTWOOD
ANNULAR RINGS
PITH
BARK
SAPWOOD
BAST
(g) CROSS SECTION THRO' TREE

Figure 13

Dry Rot

Care must always be taken to ensure that dry rot does not attack the timber. *Dry rot* is a fungoid which is continually present in the atmosphere and which will attack timber when it has a moisture content of between 20–27 per cent. To prevent attack, timber must be seasoned adequately and ventilated when fixed, to ensure the moisture content is kept sufficiently low.

Once dry rot has attacked timber it is not easy to eradicate and the timber affected should be removed and burnt, and the surrounding work adequately treated by burning to prevent re-attack on any new timber which may be used.

Timber which cannot be adequately ventilated should be protected by tar, creosote, or one of the several proprietary treatments on the market.

Cutting of Timber

Timber will be cut from the log to suit the purpose for which it is to be used as shown in Figures 13(e) and (f) and it is important to understand the results which will occur under different methods of cutting, since when timber is seasoned it will shrink and a change in shape will occur.

Defects in Timber

Since timber is a natural material certain defects can occur and the more common are given below:

(*a*) *Heart Shake, Star Shake and Cup Shake.* These are splits in the heart of the tree as shown in Figure 13 and are not usually discovered until the tree is felled. The size and depth of the crack will vary.

(*b*) *Upsetts.* A break in the grain of the wood as shown in Figure 13(d).

Preparation of Timber

Timber is often purchased in random lengths and nominal sizes, the contractor having to prepare the timber according to the size and length required.

Timber will be sawn to the required size and planed to provide a smooth finish. The planing will reduce the size of timber and certain allowances are made to provide for this. Thus the nominal size of planed timber will, in fact, be slightly greater than the actual size, $\frac{3}{32}$ inches on each planed face being allowed. Timber is referred to by the nominal size.

Timbers used for Building

The classification of softwoods and hardwoods is not related to the degree of hardness of the wood, but to the source of the tree. Softwoods come from the coniferous trees and hardwoods from the deciduous trees. Because of this classification certain trees termed softwoods are in fact of a harder texture than some of the hardwoods and the opposite is also true. A notable example of this is balsa wood, which is extremely soft, yet it is classified as a hardwood.

The following are common examples of softwoods and hardwoods used in building work:

Softwoods

Cedar. There are several varieties, the most common being red cedar. It is very durable but soft and is used for fencing, shingles, etc.

Douglas Fir. Also known as British Columbian pine and Oregon pine. Used for flooring, doors, etc.

Pine. There are several types, e.g. Pitch, Kauri, Red, Scots, Canadian White, Western White. All these timbers are used in joinery work.

Spruce. There are several types, e.g. Canadian, Common (whitewood) and Silver (sitka). Not very durable timbers. Used for constructional work, carcassing, etc.

Hardwoods

Beech. Very hard with pleasing figuring. Used for flooring.

Birch. Easily worked timber, is hard but not durable. Used for flooring and veneers.

Elm. A hard timber. Used for sills.

Mahogany. Durable and hard. Used for joinery, panelling, veneers.

There are a number of African mahoganies which have taken the place of the Honduras, e.g. Obeche.

Oak. There are several varieties available, among which are English, Japanese and American. Strong timbers and durable. Used for flooring, panelling and joinery work.

Teak. Extremely durable. Used for floors, doors, stairs, etc.

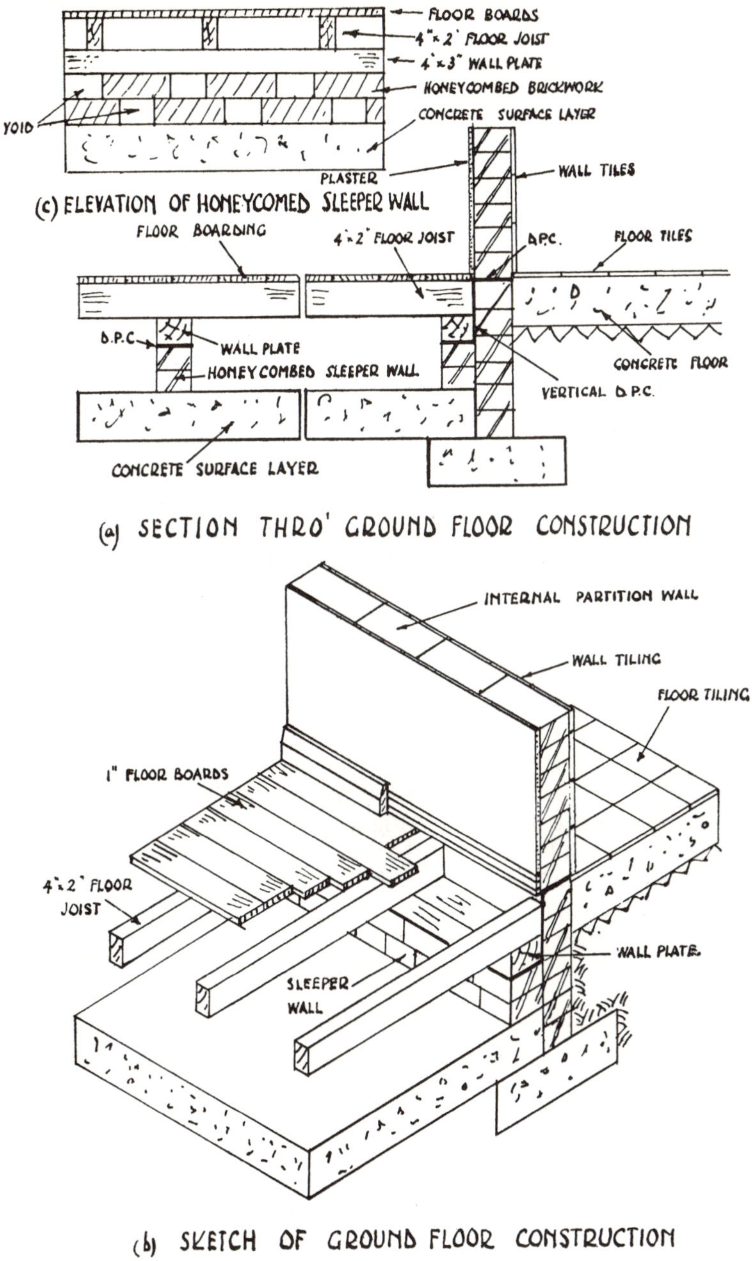
FLOOR BOARDS
4" × 2" FLOOR JOIST
4" × 3" WALL PLATE
HONEYCOMBED BRICKWORK
VOID
CONCRETE SURFACE LAYER
PLASTER
WALL TILES
(c) ELEVATION OF HONEYCOMED SLEEPER WALL
FLOOR BOARDING
4" × 2" FLOOR JOIST
D.P.C.
FLOOR TILES
D.P.C.
WALL PLATE
HONEYCOMBED SLEEPER WALL
CONCRETE FLOOR
VERTICAL D.P.C.
CONCRETE SURFACE LAYER
(a) SECTION THRO' GROUND FLOOR CONSTRUCTION
INTERNAL PARTITION WALL
WALL TILING
FLOOR TILING
1" FLOOR BOARDS
4" × 2" FLOOR JOIST
WALL PLATE.
SLEEPER WALL
(b) SKETCH OF GROUND FLOOR CONSTRUCTION

Figure 14

7 Floors

Ground Floors may be either hollow or solid. The former will consist of timber joists supported on walls with a covering of floor boarding. Solid floors are dealt with on page 42.

Floor Joists will usually be 2″ wide and of a depth suitable to the span.

Since the ground floor joists can be supported at intervals by sleeper walls, the depth of the joist can be nominal and will usually be 4″ with the joist supported at approximately 6 ft intervals.

Sleeper walls are half brick thick dwarf walls built on the surface layer, and their function is to support the ground floor joists. When placed against the main walls they will be built solid, but where independent of the main walls they will be built honeycombed. The joist should be supported on a 4″ × 3″ timber wall plate which is placed on the sleeper wall, but sometimes in housing work this is omitted. An efficient damp-proof layer should be laid on the sleeper wall under the wall plate, or if there is no wall plate, under the joists, to prevent the timber becoming wet or damp, which would encourage dry rot. See Figure 14.

The joists can be nailed to the plate and care should be taken that the top surface of the joists are in the same plane, so that the floor boards, which are fixed later, are level.

Joists adjacent to walls should be fixed with at least 1″ clearance from the wall to allow for ventilation.

Spacing of Joists

The common spacing of floor joists is 15″ centre to centre, but they can be less than this, if desired. Greater spacing is not recommended.

Floor Boards

The properties which a floor should possess are resistance to wear, resilience, thermal insulation, and suitable appearance for the building.

Wood flooring can be used in most situations, provided the type

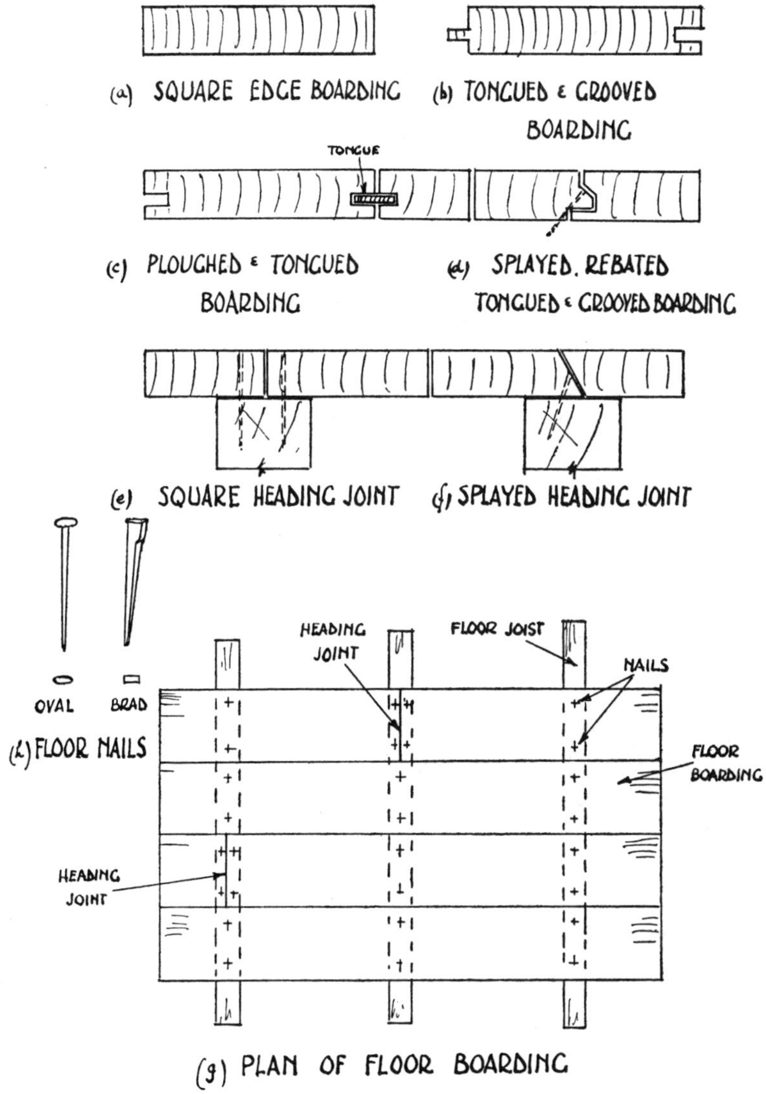

(a) SQUARE EDGE BOARDING
(b) TONGUED & GROOVED BOARDING
TONGUE
(c) PLOUGHED & TONGUED BOARDING
(d) SPLAYED, REBATED TONGUED & GROOVED BOARDING
(e) SQUARE HEADING JOINT
(f) SPLAYED HEADING JOINT
OVAL
BRAD
(h) FLOOR NAILS
HEADING JOINT
FLOOR JOIST
NAILS
FLOOR BOARDING
HEADING JOINT
(g) PLAN OF FLOOR BOARDING

Figure 15

of timber is selected carefully and its quality and moisture content controlled to suit the condition of its position.

The thickness of floor boards should be not less than ¾″ but should, preferably, be thicker. With the spacing of the joists at 15″ centres as given above the boards should be 1″ thick minimum to prevent deflection when loaded.

The cutting of timber for floor boards will be as given on page 35, which will give the most economical wear.

The jointing between the edge of the boards may be one of the following:

Square edge. The simplest form of floor boarding, which has the disadvantage of the boards tending to curl across their width as shown in Figure 15(a), particularly if not properly seasoned for their position.

Tongued and Grooved. The board will be grooved on one edge and tongued on the other as shown in Figure 15(b). This will restrict movement across the board making a more satisfactory floor.

Ploughed and Tongued. Similar to the last but both edges are grooved and a separate hardwood or metal tongue is inserted between the boards as shown in Figure 15(c).

Secret Nailed. In positions where the boards are not covered it may be necessary to hide the nailing and one method is shown in Figure 15(d).

Heading Joint. It will not always be possible to make the floor boards of the same length as the room and a joint will be necessary in the length. When this takes place the joint must be over a joist and one of two types of joint may be used as shown in Figures 15(e) and (f).

Fixing Boards. As each board is laid it will be cramped to the board adjacent to it and nailed with two oval nails, ½″ longer than the thickness of the boarding, to each joist. The use of floor brads has been superseded by oval nails in domestic work. Figure 15(h) shows both of these nails.

Solid Ground Floors

For certain types of floor finish a solid floor is essential. The advantage constructionally of a solid floor is the elimination of dry rot and other troubles normally associated with hollow joisted floors. The disadvantage is that the floor is less resilient to walk upon and may be more tiring for the user.

The solid floor will consist of a layer of concrete, which in the case of a domestic building will be the surface layer brought up to ground floor level with hardcore filling under it. The usual situations for solid floors in a domestic building are the hall and the kitchen,

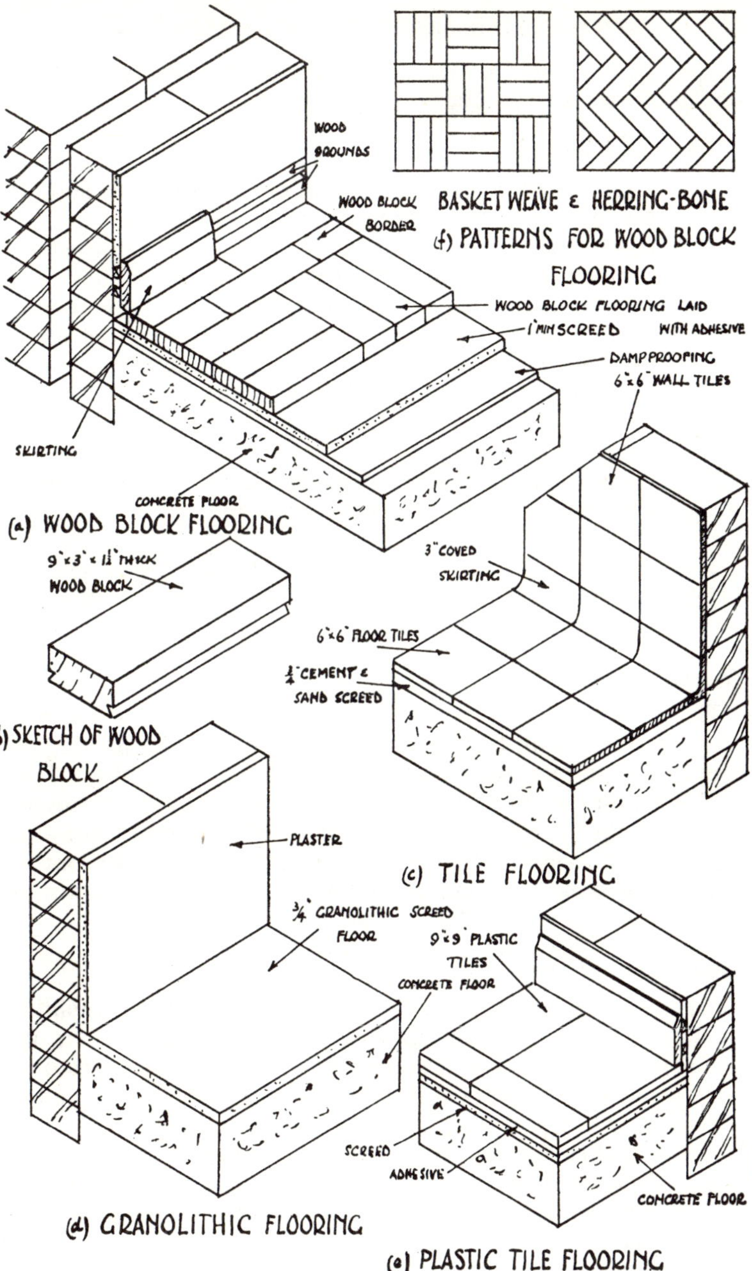
WOOD GROUNDS
WOOD BLOCK BORDER
BASKET WEAVE & HERRING-BONE
(f) PATTERNS FOR WOOD BLOCK FLOORING
WOOD BLOCK FLOORING LAID WITH ADHESIVE
1" MIN SCREED
DAMP PROOFING
6"x6" WALL TILES
SKIRTING
CONCRETE FLOOR
(a) WOOD BLOCK FLOORING
9"x3"x1½" THICK WOOD BLOCK
(b) SKETCH OF WOOD BLOCK
3" COVED SKIRTING
6"x6" FLOOR TILES
¼" CEMENT & SAND SCREED
(c) TILE FLOORING
PLASTER
¾" GRANOLITHIC SCREED FLOOR
9"x9" PLASTIC TILES
CONCRETE FLOOR
SCREED
ADHESIVE
CONCRETE FLOOR
(d) GRANOLITHIC FLOORING
(e) PLASTIC TILE FLOORING

Figure 16

but will be necessary for other rooms where wood blocks and other similar finishes are required.

Waterproof Membrane

This is the most important feature of the solid ground floor as most of the trouble affecting the finishings of these floors is attributable to a poor membrane or bad workmanship in its construction. Damp must be excluded from the floor for most types of finish and the choice, and supervision, of the application of the damp-proofing must be strictly observed. The most satisfactory waterproof membrane is asphalt, but this is expensive for domestic work. It will consist of one or two layers, applied hot, over the whole area of the concrete floor. A further layer of concrete may be placed on the asphalt if required to protect it from damage.

Plastic or bituminous sheet may also be laid on the concrete, care being taken that there is sufficient overlap and seal between the sheets.

There are several liquid applications such as bitumen and tar which can be used, but care must be taken to ensure they are correctly applied to the whole area of the floor.

Protective Screeds

The protection of the membrane is important and a screed will often be sufficient in this respect. The screed will also provide an even surface to receive the finishings, particularly those of a soft or pliable nature such as rubber, linoleum, etc., which will show every blemish in the supporting surface. The screed will consist of a fine concrete mix composed of sand and cement in the proportions of one part of cement to four or five parts of sand, applied with a trowel or wood float and brought up to an even surface.

Solid Floor Finishes

Granolithic. Where areas are to receive hard wear a granolithic paving will be suitable, although its appearance would not normally be suitable for internal domestic work. Granolithic is composed of cement and fine aggregate mortar, the aggregate being granite chippings, which will give the hard wearing quality of the finish. It will be laid as a screed, trowelled or floated to an even and fine finish. There are other applications of granolithic paving, but they will not normally come within the scope of the work being covered at this stage. See Figure 16(d).

Floor Tiling. The use of floor tiles has been common for domestic work over a long period.

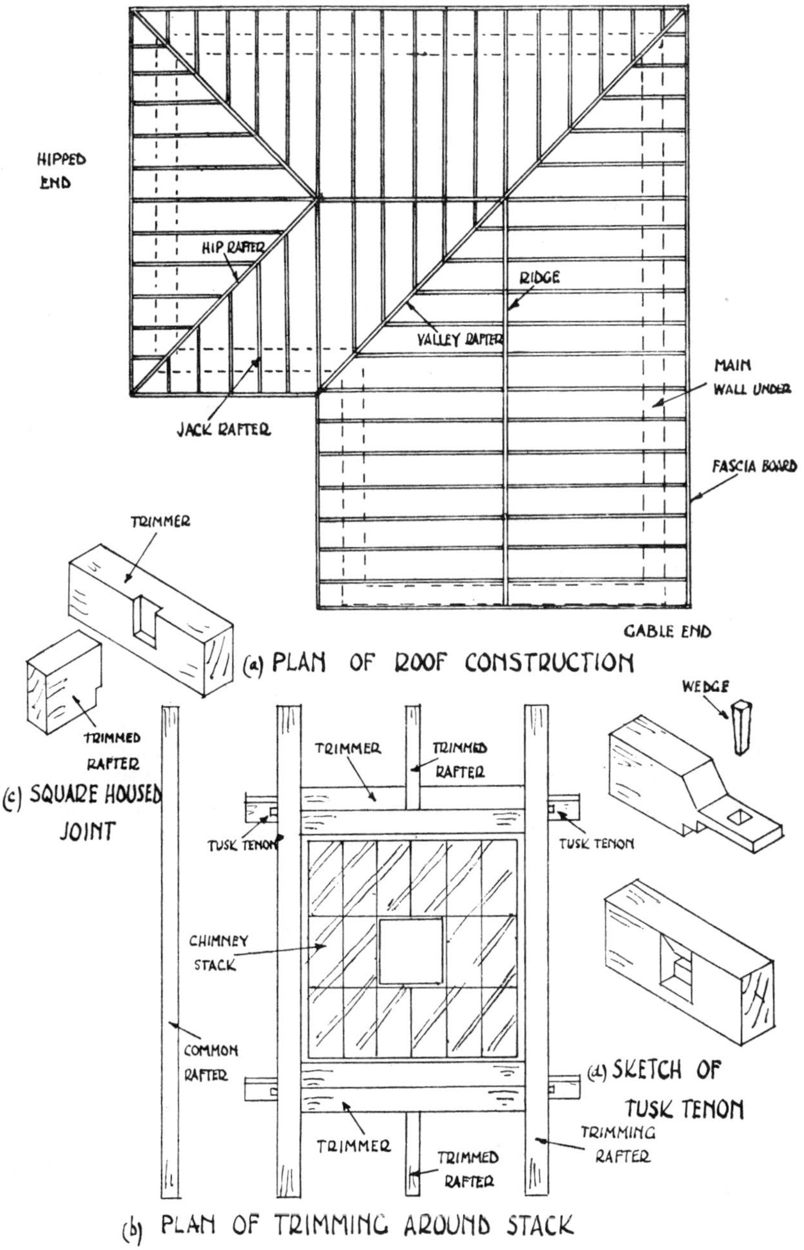
HIPPED END
HIP RAFTER
RIDGE
VALLEY RAFTER
MAIN WALL UNDER
JACK RAFTER
FASCIA BOARD
GABLE END
(a) PLAN OF ROOF CONSTRUCTION
TRIMMER
TRIMMED RAFTER
(c) SQUARE HOUSED JOINT
WEDGE
TRIMMER
TRIMMED RAFTER
TUSK TENON
TUSK TENON
CHIMNEY STACK
COMMON RAFTER
(d) SKETCH OF TUSK TENON
TRIMMER
TRIMMED RAFTER
TRIMMING RAFTER
(b) PLAN OF TRIMMING AROUND STACK

Figure 17

Quarry Tiles. These have been the most popular, particularly for kitchens, halls and entrance porches. Quarry tiles are a product manufactured in a similar manner to bricks, the materials being of finer quality. Care should be taken in the selection of a quarry tile to ensure that it is of good quality, as the main defect in this type of floor is the wearing of the top surface, which will collect the dirt and require a considerable amount of cleaning. Quarry tiles make a good impervious floor finish, but are hard and cold to the feet. They do not require a damp-proof membrane and will be fixed to the screed with cement and sand mortar and pointed.

The junction with the walls may be concealed with a cove as shown in Figure 16(c).

Terrazzo Tiles. These are good wearing and decorative tiles as an alternative to quarry tiles. They are manufactured by mixing a mortar of aggregate composed of glass, marble, ebony or other decorative hard wearing material with cement. After setting, the top surface is ground smooth to expose the aggregate, and polished.

Plastic Tiles. These have become very popular for domestic work, because of the variety of patterns available and their economic cost. They are manufactured from polyvinyl resin. These tiles will become brittle under cold conditions. See Figure 16(e).

Linoleum and Rubber. These are available in tile form and will provide a warm and resilient walking surface. They should be fixed with special adhesives and the rubber should also have a latex screed. These floorings may also be obtained in sheet form.

Wood Blocks. These are of hardwood in blocks 9″ × 3″ and 1½″ thick laid on a screed with adhesive. Various patterns are used, the most common being 'basket weave' and 'herring-bone' as shown in Figures 16(a), (b) and (f).

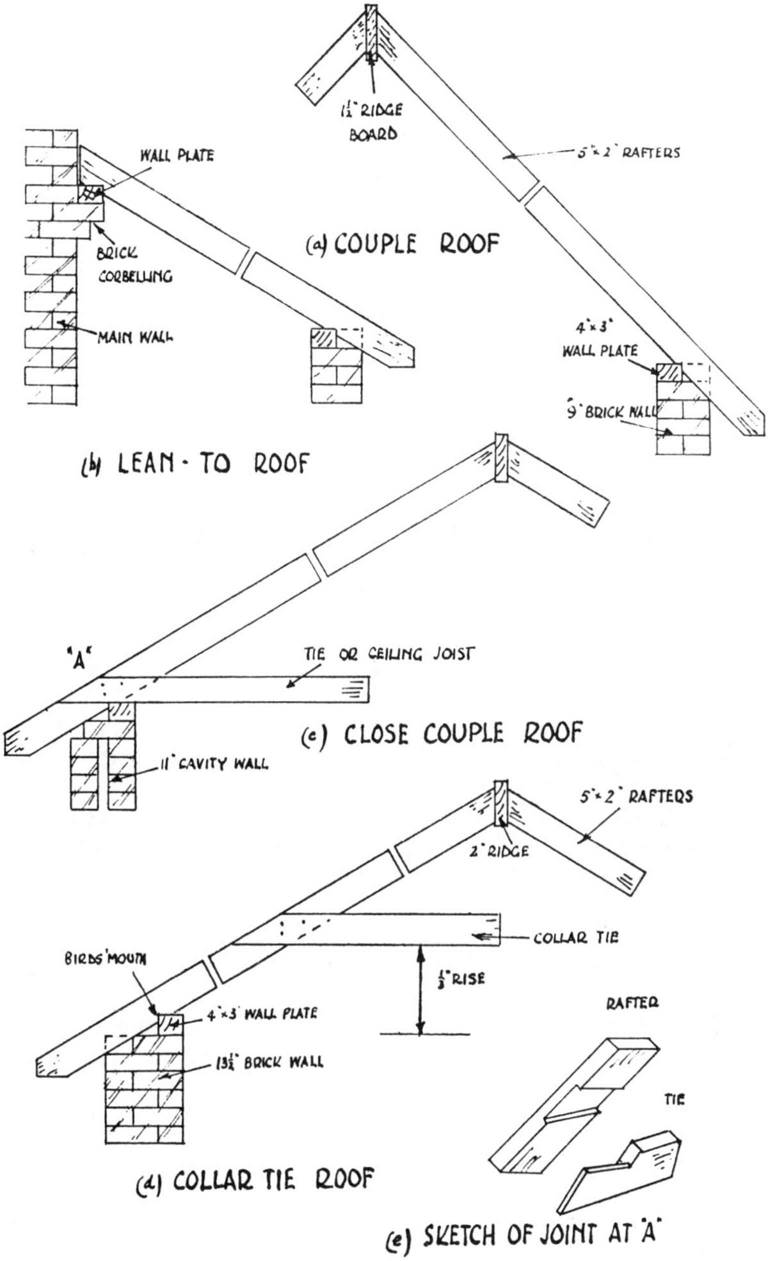
1½" RIDGE BOARD
5"×2" RAFTERS
WALL PLATE
BRICK CORBELLING
MAIN WALL
(a) COUPLE ROOF
4"×3" WALL PLATE
9" BRICK WALL
(b) LEAN-TO ROOF
"A"
TIE OR CEILING JOIST
(c) CLOSE COUPLE ROOF
11" CAVITY WALL
5"×2" RAFTERS
2" RIDGE
COLLAR TIE
BIRDS'MOUTH
⅓ RISE
4"×3" WALL PLATE
13½" BRICK WALL
RAFTER
TIE
(d) COLLAR TIE ROOF
(e) SKETCH OF JOINT AT "A"

Figure 18

8 Pitched Roofs

Roofs

Pitched roofs may be hipped or gabled at their ends as shown in Figure 17(a). The common terms related to roofs are also given.

Lean-to Roof

The simplest form of roof with support at one side on a main structural wall and at the other side on an independent wall, as shown in Figure 18(b). The pitch will be regulated by the type of covering.

Couple Roofs

The usual form of roof is that which is pitched from both sides and the simplest form of these is the *couple roof*, as shown in Figure 18(a). The disadvantage of this roof is that the load from the coverings tends to spread the feet of the rafters. To overcome this a tie member can be fixed across the feet of the rafters as shown in Figure 18(c). The tie members can also be used as ceiling joists where the room below is to be ceiled. To obtain more headroom the tie member may be raised as shown in Figure 18(d) to form a *collar tie roof*.

Rafters

Rafters are similar to floor joists in their function, the loads being generally less and the method of support different.

They will be pitched to suit the type of roof covering and the spacing will be 15″ centre to centre. The width of the rafters will be 2″, the depth being related to the clear span.

At the top the rafters will be supported and spiked to a ridge board, which is provided to allow easier fixing, while at the foot the rafter will be supported on a wall plate, over which it will be birdsmouthed and spiked.

As the span of the roof increases the length of the rafter will increase and will require additional depth or intermediate support, the latter being provided by the addition of one or more purlins.

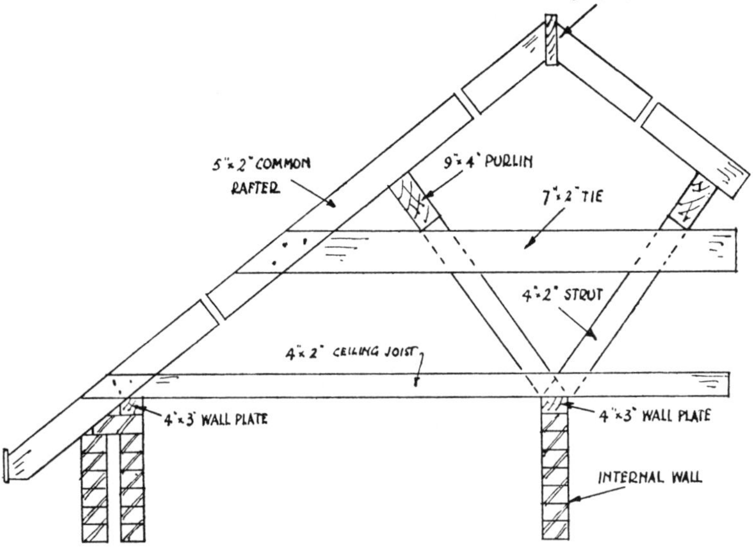

(a) SECTION THRO' ROOF

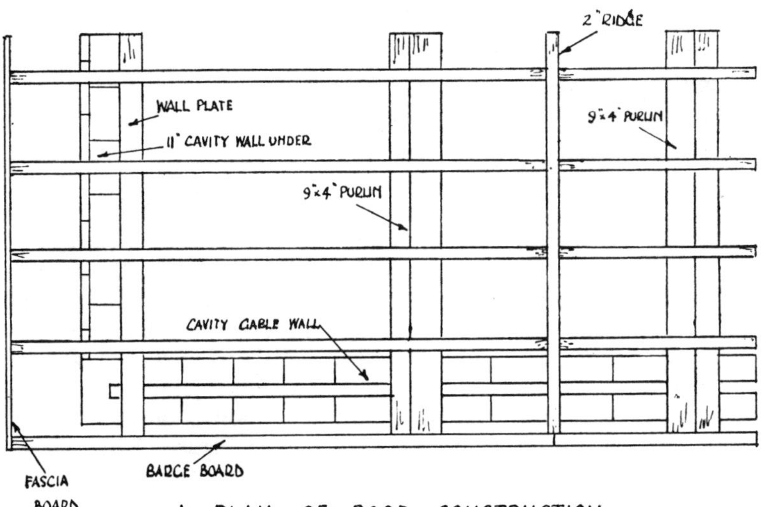

(b) PLAN OF ROOF CONSTRUCTION

Figure 19

The purlin may be supported on a gable wall or where hipped roofs are concerned, will be supported by struts from internal walls below, as shown in Figure 19(a). For larger spans of roof a truss in steel or timber may be required.

Gable

Where the external wall is built to the apex of the roof, instead of a hipped end, the roof covering will finish with a verge. The ridge board will be supported on the gable brickwork, see Figure 19(b). The rafters will commence one inch from the inside of the gable to allow for ventilation. Alternatively, a rafter may be placed on either side of the gable wall, the ridge, purlin and wall plate being allowed to project beyond the outer face of the gable brickwork to support them. This will provide an overhanging verge giving a better appearance to the end of the house. A barge board may be fixed to the end of the ridge, purlin and wall plate to protect them.

Hips and Valleys

The rafters will be shortened because of the hip and valley rafters; these shorter rafters are known as *jack rafters*. They will be cut to suit the angle of the hip or valley. The hip and valley rafters will be 2″ wide and of sufficient depth to receive the jack rafters and also be of sufficient strength to carry the load of the jack rafters. The top of the hip rafter should be bevelled to receive the battens of the roof coverings.

Trimming of Openings in Roof

Where a chimney stack passes through a roof, the timbers must not be supported on it, because of fire risk. It will be necessary to trim the timbers around the opening. The rafters adjacent to the opening, called *trimming rafters*, should be increased in size, and since they cannot be made deeper must be increased in width by $\frac{1}{2}''$ or 1″. The rafters which are reduced in length because of the opening are termed *trimmed rafters* and will be supported on the *trimmer*, which will span between the trimming rafters. The trimmer will be increased in size as already stated for the trimming rafters, because of the additional load it must carry, and the joint between the trimmer and trimming rafter will be a tusk tenon as shown in Figure 17(d). A sketch of these members is shown in Figure 17(b).

Ceiling Joists

Unless the roof space of a building is being used for room space, the ceiling joists do not require to be as deep as for floor joists. In

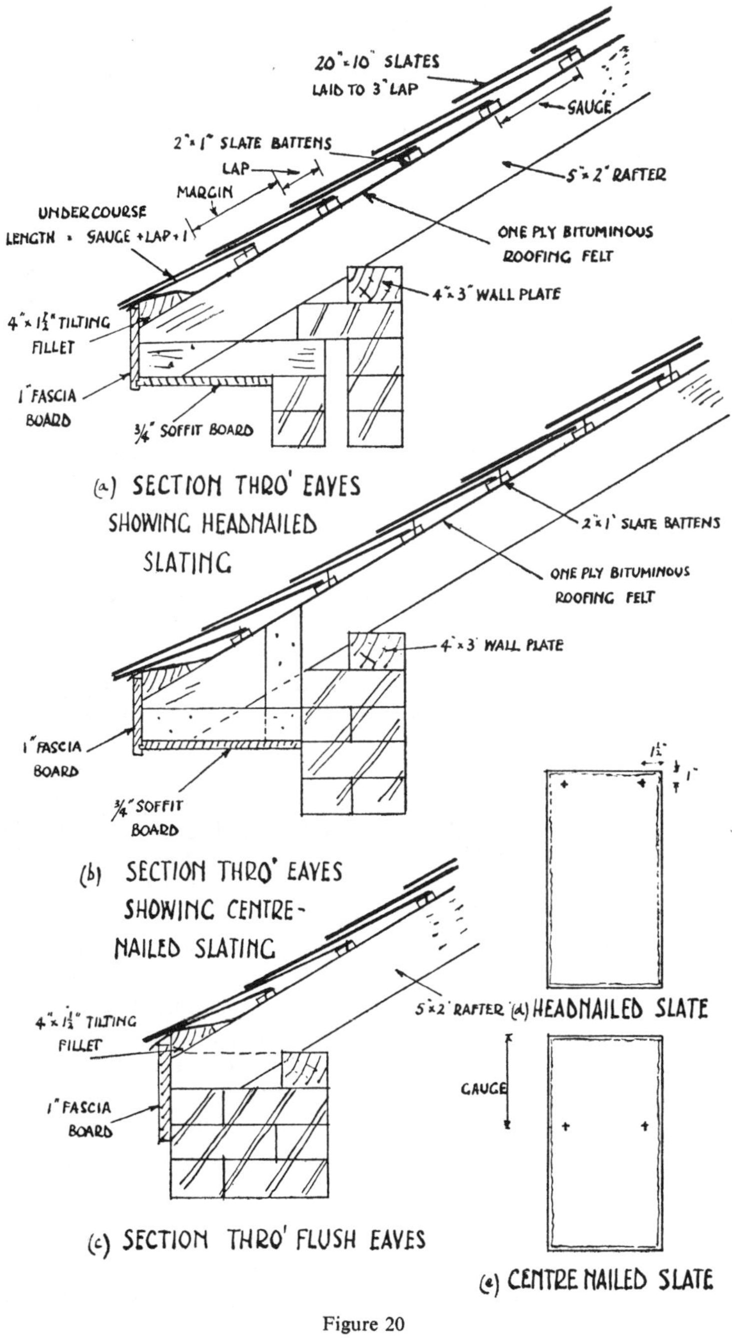
20" × 10" SLATES
LAID TO 3" LAP
GAUGE
2" × 1" SLATE BATTENS
LAP
MARGIN
5" × 2" RAFTER
UNDERCOURSE
LENGTH = GAUGE + LAP + 1
ONE PLY BITUMINOUS
ROOFING FELT
4" × 3" WALL PLATE
4" × 1½" TILTING
FILLET
1" FASCIA
BOARD
¾" SOFFIT BOARD
(a) SECTION THRO' EAVES
SHOWING HEADNAILED
SLATING
2" × 1" SLATE BATTENS
ONE PLY BITUMINOUS
ROOFING FELT
4" × 3" WALL PLATE
1" FASCIA
BOARD
¾" SOFFIT
BOARD
(b) SECTION THRO' EAVES
SHOWING CENTRE-
NAILED SLATING
1½"
1"
4" × 1½" TILTING
FILLET
1" FASCIA
BOARD
5" × 2" RAFTER
(d) HEADNAILED SLATE
GAUGE
(c) SECTION THRO' FLUSH EAVES
(e) CENTRE NAILED SLATE

Figure 20

consequence a size of 4″ × 2″ is normally sufficient. Care must be taken in fixing ceiling joists that the underside of the joists are level, as any variation between one joist and another is reflected in the plasterboard ceiling.

Roof Coverings

The traditional forms of roof covering are slate and tile, the latter being more commonly used today.

Slating

Slates are quarried and obtained from a number of areas in the country, the usual sources being Westmorland, Wales and Cornwall. The blocks of slate are cut out of the quarry face and split into sheets, along their natural cleavage planes. The quality of the slate is related to its soundness and thickness. A good quality slate is one which is sound and thin, thus reducing the weight on the roof and at the same time being completely waterproof.

Slates are cut into sizes which are suitable for handling. At one time they were classified by name, but this has been superseded by size.

The usual sizes are 16″ × 8″, 18″ × 9″, 20″ × 10″ and 22″ × 12″, but other sizes between these can be obtained. Slates may be purchased by the thousand of regular size, or by weight, if of random size.

The slates will be delivered to the site without nail holes and the roofer will cut the holes to suit the architect's fixing requirements. Slates can be head nailed or centre nailed, see Figures 20(d) and (e). Generally, the larger slates will be centre nailed to reduce the wind leverage. The disadvantage of centre nailing is that the nail hole is not covered as adequately as in head nailing.

Figures 20(a) and (b) show the method of fixing regular size slates both centre and head nailed and it will be seen that in both instances the pattern of covering is the same, viz. two slate thicknesses everywhere, and three slate thicknesses at each batten. This will ensure a dry roof even if one slate should become dislodged or broken.

Slate nails should be non-ferrous or adequately protected against rusting, and copper, composition or galvanised nails should be used. The lowest pitch for slating is 27°.

Eaves

In roofing work the coverings commence at the eaves and in order to provide two courses of slates a shorter underslate will be necessary to commence the coverings. The underslate will be sup-

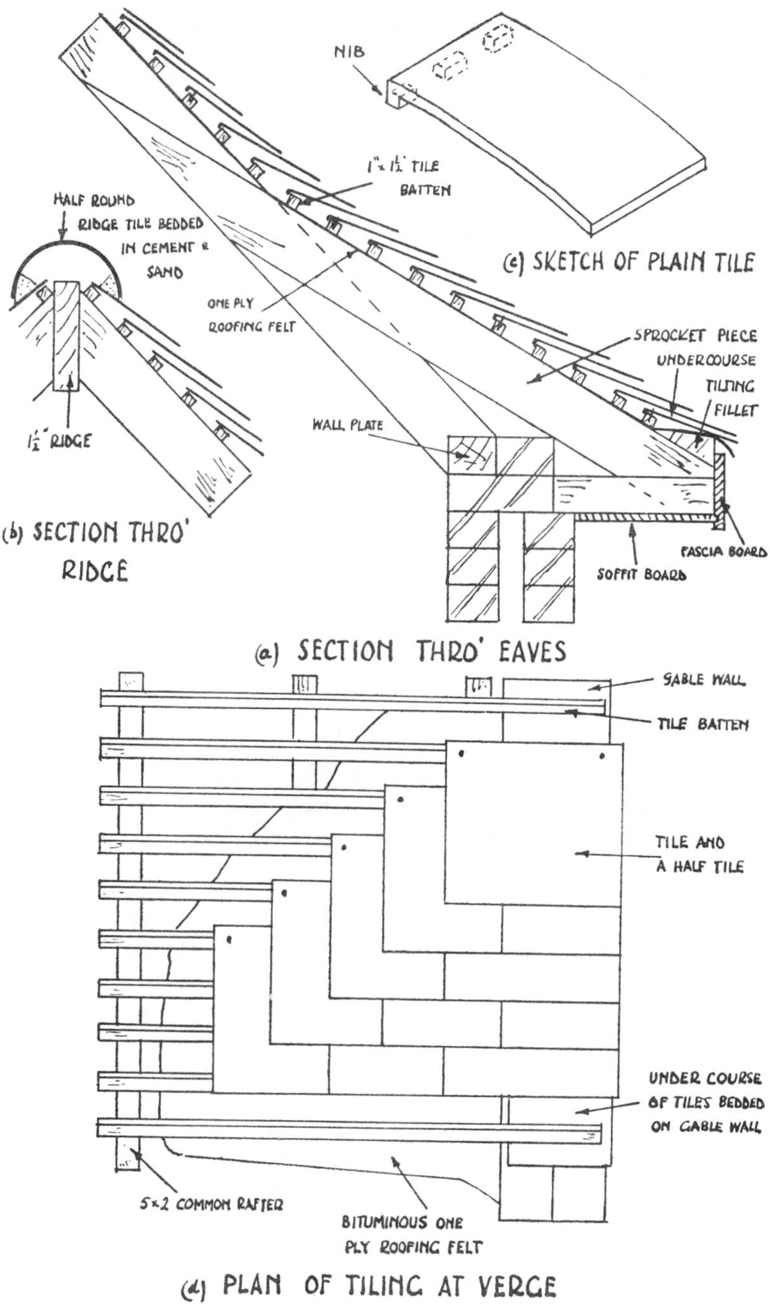
NIB
1" × 1½" TILE BATTEN
(c) SKETCH OF PLAIN TILE
HALF ROUND RIDGE TILE BEDDED IN CEMENT & SAND
ONE PLY ROOFING FELT
1½" RIDGE
(b) SECTION THRO' RIDGE
SPROCKET PIECE
UNDERCOURSE
TILTING FILLET
WALL PLATE
FASCIA BOARD
SOFFIT BOARD
(a) SECTION THRO' EAVES
GABLE WALL
TILE BATTEN
TILE AND A HALF TILE
UNDER COURSE OF TILES BEDDED ON GABLE WALL
5×2 COMMON RAFTER
BITUMINOUS ONE PLY ROOFING FELT
(d) PLAN OF TILING AT VERGE

Figure 21

ported on a tilting fillet. There are several forms in which the eaves can be constructed, as shown in Figure 20.

Verge

The finish of the roof coverings at the verge will require special treatment to provide a bond for the slates. To overcome the problem of half a slate every other course, which will tend to break away from the roof, a slate and half slate will be used similar to that shown in Figure 21(d) which shows the finish with a flush verge. The verge should be tilted to throw the water on to the roof.

Ridge

The last row of roof coverings to be laid at the ridge will rarely be a full slate or tile, in consequence, the covering will be shorter as shown in Figure 21(b), which also shows the covering to the ridge. Ridge coverings may be slate, lead, or burnt clay half-round ridge tiles.

Abutment

The finish at the abutment will be with lead or copper soakers and stepped flashings as shown in Figure 23.

Tiling

Plain tiles are $10\frac{1}{2}'' \times 6\frac{1}{2}''$ and $\frac{1}{2}''$ thick, manufactured from clay in a similar manner to bricks. Tiles being smaller than slates, they are curved on their bed to overcome any capillarity, which would tend to take place if the tiles were made flat. The curve and shape of tiles can be seen in Figure 21(c). Each tile will have two nail holes, although the nibs will provide the support for most of the tiles on normal pitched roofs. In domestic work the first and last courses will be nailed and every fourth or fifth in between.

Nails will be of copper, composition or galvanised to prevent failure by rusting.

The least pitch for tiles is 35°.

Eaves, verge, ridge, and abutment will be of similar finish to that given for slates, as shown in Figure 21.

Hips. An external intersection of roof surfaces form a hip and in slating the intersection can be made watertight by lead or copper soakers laid with the slating. Alternatively, a half-round hip tile can be bedded in cement and sand on the slates.

For tiling the hip can be finished as described for slating, or, as is more common, a purpose-made hip tile is used in conjunction with the tiling, as shown in Figure 22(b).

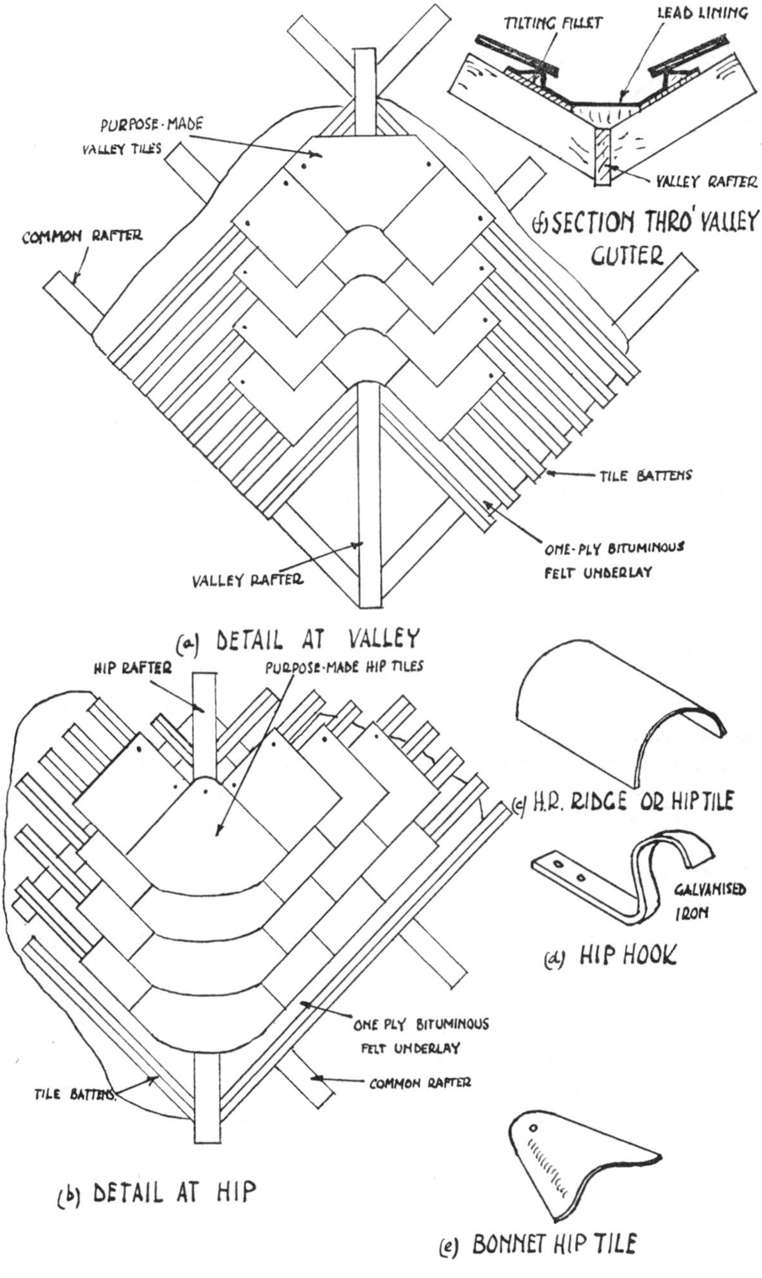
PURPOSE-MADE
VALLEY TILES
COMMON RAFTER
TILTING FILLET
LEAD LINING
VALLEY RAFTER
(f) SECTION THRO' VALLEY
GUTTER
TILE BATTENS
ONE-PLY BITUMINOUS
FELT UNDERLAY
VALLEY RAFTER
(a) DETAIL AT VALLEY
HIP RAFTER
PURPOSE-MADE HIP TILES
(c) H.R. RIDGE OR HIPTILE
GALVANISED
IRON
(d) HIP HOOK
ONE PLY BITUMINOUS
FELT UNDERLAY
COMMON RAFTER
TILE BATTENS
(b) DETAIL AT HIP
(e) BONNET HIP TILE

Figure 22

Valleys. An internal intersection of roof surfaces form a valley and can be made weathertight by lead or copper soakers or by a lead or copper lined gutter, as shown in Figure 22(f).

For tiling the same methods may be used or a purpose-made valley tile can be used in conjunction with the tiling, as shown in Figure 22(a).

Battens

Both slates and tiles will be fixed on sawn timber battens, the size being 2″ × 1″ and 1″ × ¾″ respectively, and nailed to the rafters, or roof boarding, with galvanised or composition nails.

When roof boarding is fixed under the roof covering counter-battens will be required under the slate or tile battens to raise them, allowing any rain or snow which may penetrate under the coverings to flow unimpeded to the gutter at the foot of the rafters.

Every slate will be nailed to the battens but, as already stated, tiles will be nailed only every fourth or fifth course. Where the pitch is steep or vertical, however, each tile will be nailed.

Felt

It is required that all domestic roofs should have a complete covering of roofing felt. Roofing felt will be one-ply bituminous felt purchased in rolls and will be fixed to the rafters or roof boarding, by the slate or tile battens. The felt, which is called 'sarking' in some areas, will prevent driving rain or snow penetrating under the roof covering and settling in the roof space, causing eventual damage to the rooms below. The felt may be seen in Figures 20, 21 and 22.

Weathering of Opening

The penetration of a stack through a roof creates a weakness in the roof from a weathering point of view. Adequate protection must be given to ensure that rain and snow do not penetrate to the inside of the building. Figure 23 shows the method of weathering, using lead. A gutter is provided at the top of the opening to direct the collection of rainwater to the side of the opening. An apron forms the protection at the lower portion of the opening and soakers and a stepped flashing at the sides. The weathering can be carried out in a similar manner with copper, zinc or aluminium.

Soakers will be provided one for each slate or tile course abutting the stack, a stepped flashing will cover the vertical upstand of the soaker and make the opening weatherproof.

When the stack penetrates the roof at the apex the gutter will be omitted and two aprons provided.

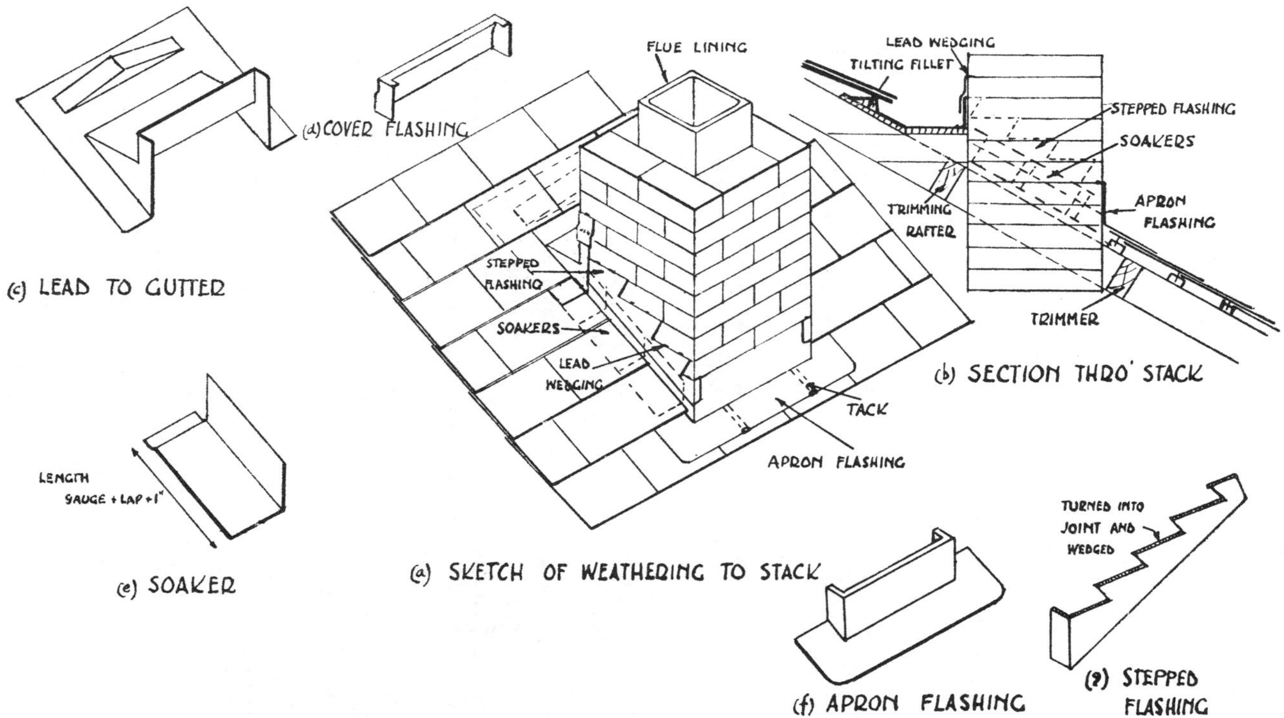
(c) LEAD TO GUTTER
(d) COVER FLASHING
FLUE LINING
LEAD WEDGING
TILTING FILLET
STEPPED FLASHING
SOAKERS
APRON FLASHING
TRIMMING RAFTER
TRIMMER
(b) SECTION THRO' STACK
STEPPED FLASHING
SOAKERS
LEAD WEDGING
TACK
APRON FLASHING
LENGTH
GAUGE + LAP + 1"
(e) SOAKER
(a) SKETCH OF WEATHERING TO STACK
TURNED INTO JOINT AND WEDGED
(f) APRON FLASHING
(g) STEPPED FLASHING

Figure 23

9 Flat Roofs in Timber

WHEREAS the pitched roof provides a good protection for a building because of its umbrella action in quickly removing rainwater, the roof space formed by the pitch of the roof is often of little value to the building. The flat roof utilises the full height of the building.

The roof joists will be supported on the brickwork at both ends and by any intermediate walls. The depth of the joist must be sufficient to enable it to span the rooms below without sagging, while the width will normally be 2″ which will allow ample space for fixing boarding. The span is the distance between the support of the joist.

It should be noted that the strength of a timber joist is not in its width but in its depth, therefore, it is necessary to increase the depth to increase the strength.

A rule of thumb method of determining the depth of a timber joist can be obtained by taking the space in feet, call it inches, and divide by two.

The fixing of the joists will be by bedding in the walls, care being taken that the tops of the joists are level to receive the boarding. When plasterboard is to be used on the underside, the joists should be of equal depth, so that variations in the depth of joists are not reflected in the fixing of the plasterboard.

The flat roof must be provided with a fall to remove the rainwater as quickly as possible. The direction of the fall will be determined by the position of the roof in relation to other buildings, or by its shape or size.

To obtain the fall firring pieces will be fixed to the top of the roof joists as shown in Figure 24(e), and upon which the roof boarding is nailed. Roof boarding should be $\frac{3}{4}$″ to 1″ in thickness and be tongued and grooved, and should, if possible, be laid with the fall, but where this is not possible, the boarding should be laid diagonally. In both cases this will direct the water into the gutter, even if the boarding should curl. Each board will be nailed with two nails at each joist. The boarding should be cleaned off to remove blemishes and all nails punched well below the surface. To further cushion the covering one-ply bituminous felt sheeting should be laid on the boarding,

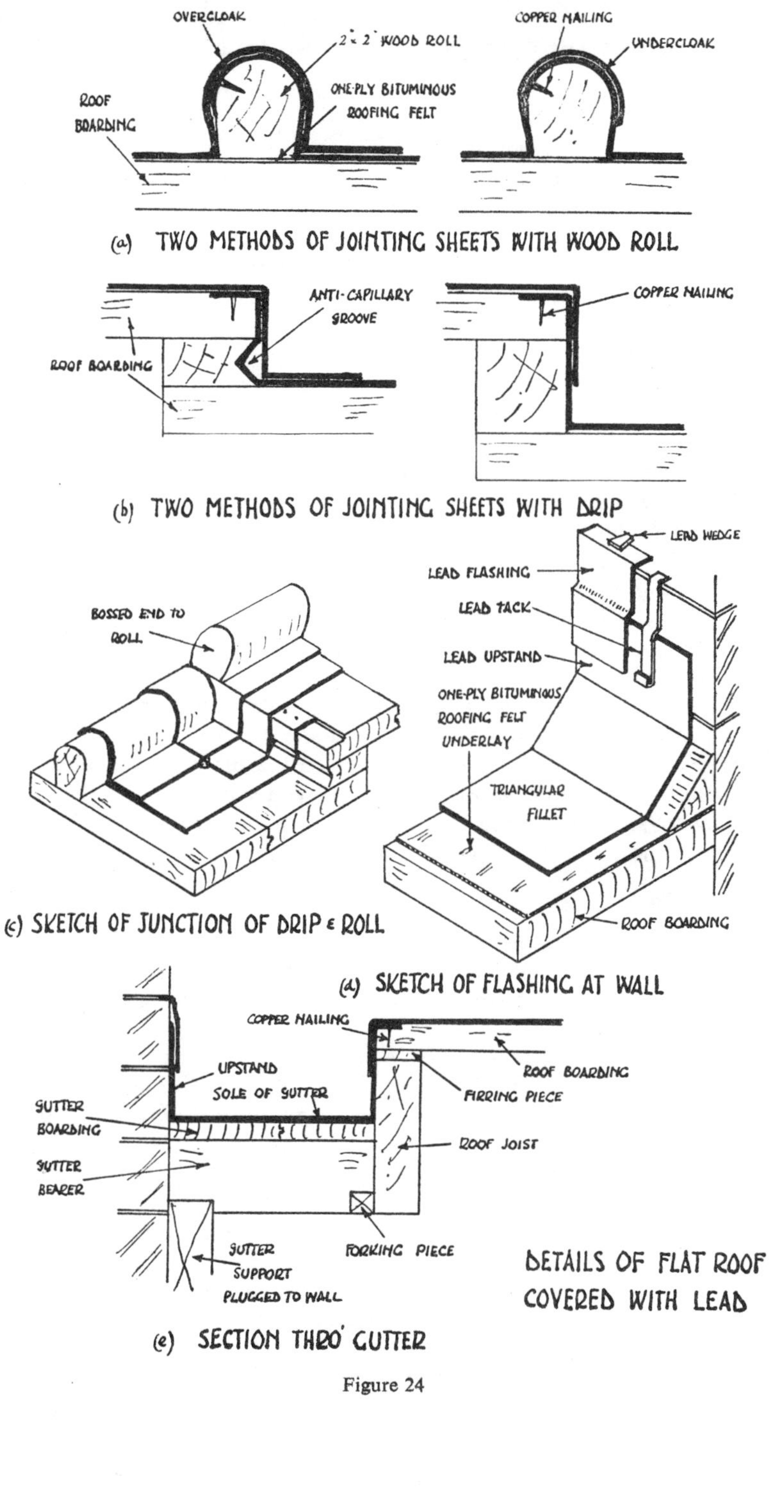
OVERCLOAK
2" x 2" WOOD ROLL
ONE-PLY BITUMINOUS ROOFING FELT
ROOF BOARDING
COPPER NAILING
UNDERCLOAK
(a) TWO METHODS OF JOINTING SHEETS WITH WOOD ROLL
ANTI-CAPILLARY GROOVE
ROOF BOARDING
COPPER NAILING
(b) TWO METHODS OF JOINTING SHEETS WITH DRIP
BOSSED END TO ROLL
(c) SKETCH OF JUNCTION OF DRIP & ROLL
LEAD WEDGE
LEAD FLASHING
LEAD TACK
LEAD UPSTAND
ONE-PLY BITUMINOUS ROOFING FELT UNDERLAY
TRIANGULAR FILLET
ROOF BOARDING
(d) SKETCH OF FLASHING AT WALL
COPPER NAILING
UPSTAND
SOLE OF GUTTER
GUTTER BOARDING
GUTTER BEARER
ROOF BOARDING
FIRRING PIECE
ROOF JOIST
GUTTER SUPPORT PLUGGED TO WALL
FORKING PIECE
(e) SECTION THRO' GUTTER
DETAILS OF FLAT ROOF COVERED WITH LEAD

Figure 24

as any blemish in the timber may cause uneven wear in the covering.

With the increased emphasis on insulation of buildings against heat losses and, in particular, where flat roofs or open type pitched roofs are concerned, wood wool slabs or straw board are used instead of timber as the support for the coverings.

Bituminous felt should be used as the base for the metal coverings, to prevent staining on the ceiling below due to condensation.

Lead Covering

Lead is manufactured by pouring molten lead on to prepared sand beds thus forming cast lead. This is rolled until the required thickness is obtained, and is termed milled lead, which is used for building purposes. It is manufactured in rolls from 33′ to 36′ long and up to 9′ 0″ in width and is classified by its weight per foot super. It can be obtained from 3 lb to 10 lb per foot super, but for constructional purposes will range from 3 lb to 6 lb. Lead moves appreciably under temperature change and this affects the size of sheet and the method of fixing. Sheets will be cut from the rolls, the width of the roll being the length of the sheet. Sheets will be approximately 3′ 0″ × 8′ 0″.

If lead is fixed rigidly on all edges the sheet would rise in the centre when warm and stretch when under very cold conditions. This movement would eventually fracture the material and allow the water to penetrate into the structure. The sheets should, therefore, be fixed on two adjacent edges only. The fall necessary for lead will be a minimum of $1\frac{1}{2}$″ in 10′.

Jointing Sheet Lead

To joint the sheet on the edge which is parallel to the fall a roll is used. This can be either a solid or hollow roll. The *solid roll* consists of a timber core fixed at 2′ 0″ centres as shown in Figure 24(a), around which the lead is dressed. Two types of the roll are shown, the variation being the difference in custom between one part of the country and another. The roll fixes the edge of one sheet and allows the edge of the adjacent sheet to move freely. The *hollow roll* acts in the same manner, but both edges of the sheet are free to move. Care must be taken in this case that persons walking on the roof do not step on the rolls.

The jointing of the sheets across the fall will be by a *drip* positioned at approximately 7′ 6″ centres according to the length of the sheet. Again two methods of jointing the lead can be used according to custom, as shown in Figure 24(b). The intersection of the roll and drip

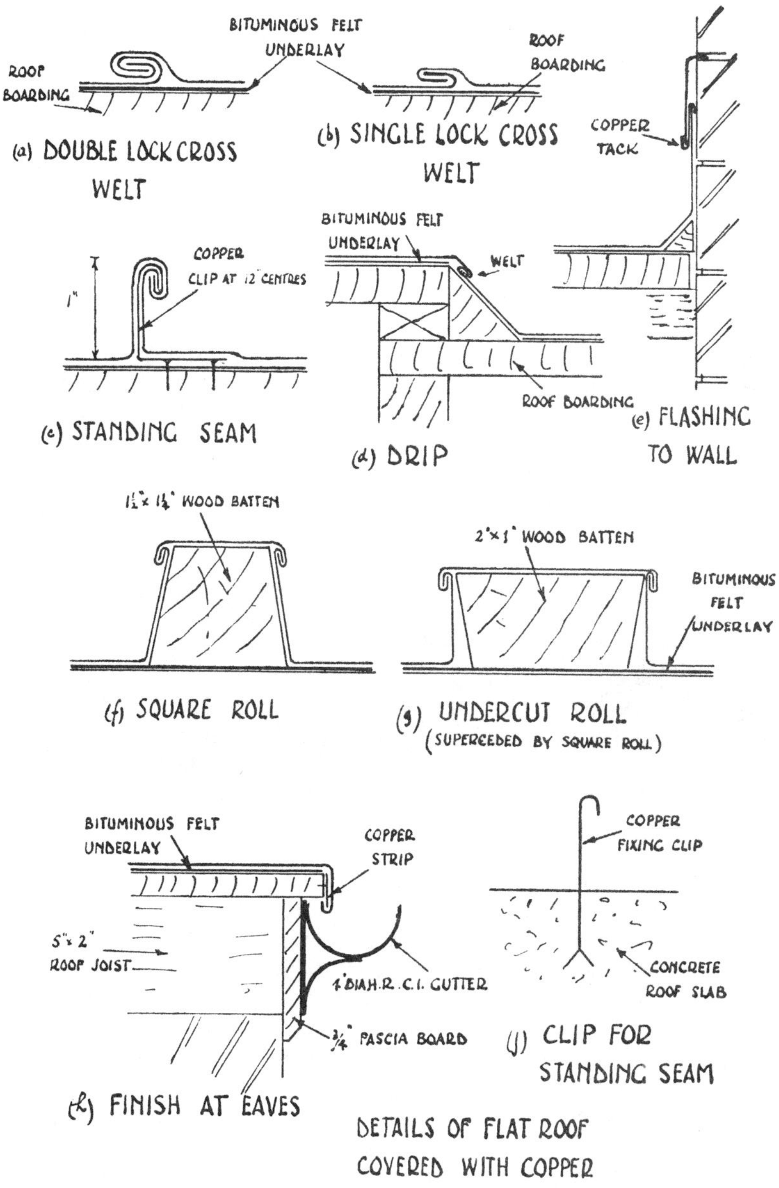
BITUMINOUS FELT UNDERLAY
ROOF BOARDING
ROOF BOARDING
(a) DOUBLE LOCK CROSS WELT
(b) SINGLE LOCK CROSS WELT
COPPER TACK
1"
COPPER CLIP AT 12" CENTRES
BITUMINOUS FELT UNDERLAY
WELT
ROOF BOARDING
(c) STANDING SEAM
(d) DRIP
(e) FLASHING TO WALL
1½" x 1½" WOOD BATTEN
2" x 1" WOOD BATTEN
BITUMINOUS FELT UNDERLAY
(f) SQUARE ROLL
(g) UNDERCUT ROLL (SUPERCEDED BY SQUARE ROLL)
BITUMINOUS FELT UNDERLAY
COPPER STRIP
5" x 2" ROOF JOIST
4" DIAM. R.C.I. GUTTER
¾" FASCIA BOARD
COPPER FIXING CLIP
CONCRETE ROOF SLAB
(h) FINISH AT EAVES
(j) CLIP FOR STANDING SEAM
DETAILS OF FLAT ROOF COVERED WITH COPPER

Figure 25

is shown in Figure 24(c). The method of providing a flashing at the wall to cover the upstand of the sheet lead is given in Figure 24(d). The jointing of the lead in gutters will follow the same form as the sheet lead as shown in Figure 24(e), a drip being necessary at the jointing of adjacent sheets.

All lead sheeting should be fixed with copper nails.

Copper Covering

Hot roll copper sheets are obtainable in lengths of 12′ but also in standard size sheets of 4′ 0″ × 2′ 0″ and 6′ 0″ × 2′ 0″ in the following gauges: 24 S.W.G. having a weight of 16 oz per sq. ft and a thickness of 0·022″; 23 S.W.G. having a weight of 19 oz per sq. ft and a thickness of 0·024″.

The fall necessary for copper will be a minimum of 2″ in 10′.

Copper does not move as freely under temperature variation as lead but, even so, it should not be fixed rigidly. Copper hardens when it is dressed and in consequence the labours on the material should be kept to a minimum. Much of the preparation will be carried out on the bench at ground level.

Joints between sheets parallel to the fall will be either rolls as shown in Figures 25 (f) and (g), or standing seams as shown in Figure 25(c), the latter being used where the passage of persons on the roof is restricted. Joints across the fall will be made with welts or drips as shown in Figures 25(a), (b) and (d). The welts will be secured with clips ¾″ wide at 2′ 6″ centres.

Brass screws or copper nails only should be used in conjunction with copper, because of electrolytic action between the metals, if ferrous nails are used. Figures 25(e) and (h) show the finish at the wall and eaves.

Bituminous Felt Roofing

The simplest type of covering and the least expensive for domestic work. This roofing should be laid on an even surface with an underlay of felt to remove any blemishes in the timber boarding. The felt consists of a sheet of fibre impregnated with bitumen formed into rolls 36″ wide and 12 yd long and can be from one- to three-ply in thickness. The one-ply is normally suitable for underlay or less important work, while the three-ply provides a very satisfactory main roof covering. The felt will be laid on the roof with cold adhesive, an upstand being provided at the junction with walls, etc. The upstand is covered with a flashing, which is tucked into the joint above the top of the upstand as shown in Figures 26(c) and (d). All sheets should be lapped 3″ to provide a sound joint.

The fall for bituminous felt is 2″ in 10′.

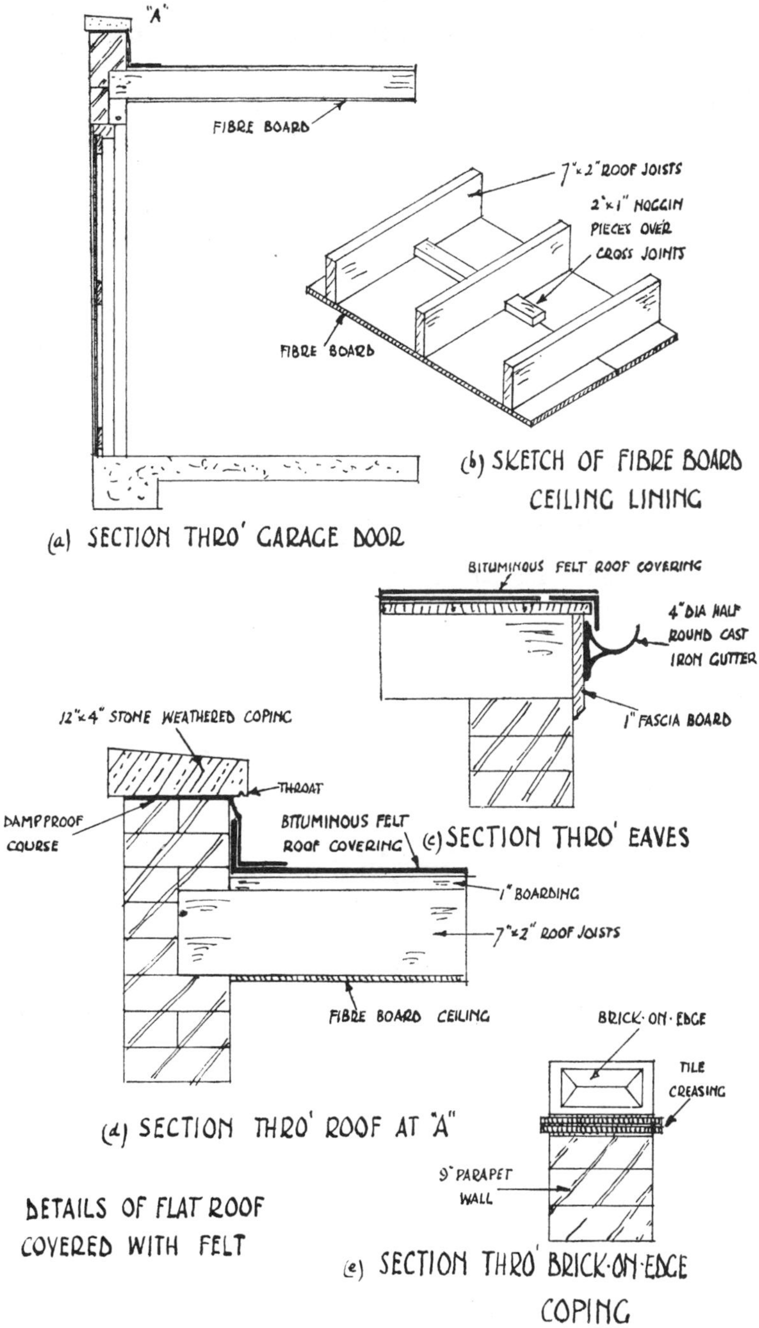

"A"
FIBRE BOARD
(a) SECTION THRO' GARAGE DOOR
7"×2" ROOF JOISTS
2"×1" NOGGIN PIECES OVER CROSS JOINTS
FIBRE BOARD
(b) SKETCH OF FIBRE BOARD CEILING LINING
BITUMINOUS FELT ROOF COVERING
4" DIA HALF ROUND CAST IRON GUTTER
1" FASCIA BOARD
(c) SECTION THRO' EAVES
12"×4" STONE WEATHERED COPING
THROAT
DAMP PROOF COURSE
BITUMINOUS FELT ROOF COVERING
1" BOARDING
7"×2" ROOF JOISTS
FIBRE BOARD CEILING
(d) SECTION THRO' ROOF AT "A"
BRICK-ON-EDGE
TILE CREASING
9" PARAPET WALL
(e) SECTION THRO' BRICK-ON-EDGE COPING
DETAILS OF FLAT ROOF COVERED WITH FELT

Figure 26

Asphalt

Asphalt comes from two sources, natural lake asphalt from Trinidad and rock asphalt from a number of places in Europe and America. Most asphalt will be manufactured from the rock, being crushed and mixed with bitumen to provide the high quality product on the market. This type of roof is not used often for domestic work because of the cost of application in relation to the size of work, but where larger buildings are to be built, asphalt will provide a lasting and competitive finish, provided it is protected against temperature change, by covering with a layer of chippings or water.

Asphalt will be laid in two or three layers according to its position. For roofs two layers will be suitable, care being taken to ensure that all angles are properly formed with an adequate fillet. Joints too, between adjoining areas of the asphalt, should be lapped to ensure that no failure can take place. When asphalt is laid on a timber base it should be provided with a bitumen felt underlay, to allow the asphalt to move independently of the roof, and also to protect the timber from the hot asphalt when laying.

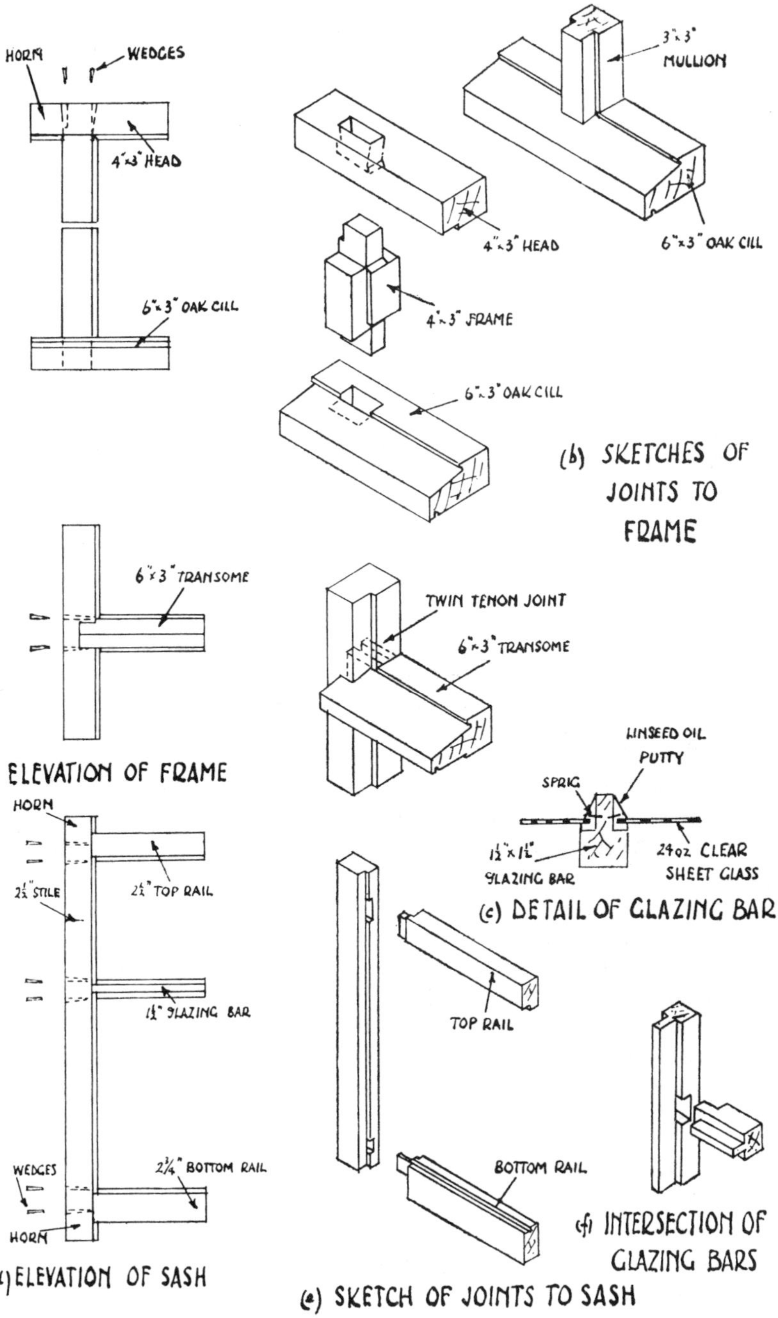
HORN
WEDGES
4"x3" HEAD
6"x3" OAK CILL
3"x3" MULLION
4"x3" HEAD
6"x3" OAK CILL
4"x3" FRAME
6"x3" OAK CILL
(b) SKETCHES OF JOINTS TO FRAME
6"x3" TRANSOME
(a) ELEVATION OF FRAME
TWIN TENON JOINT
6"x3" TRANSOME
LINSEED OIL PUTTY
SPRIG
1½"x1½" GLAZING BAR
24oz CLEAR SHEET GLASS
(c) DETAIL OF GLAZING BAR
HORN
2½" TOP RAIL
2½" STILE
1½" GLAZING BAR
WEDGES
2¾" BOTTOM RAIL
HORN
(d) ELEVATION OF SASH
TOP RAIL
BOTTOM RAIL
(f) INTERSECTION OF GLAZING BARS
(e) SKETCH OF JOINTS TO SASH

Figure 27

10 Windows

Timber Frames

The fixing of timber frames for doors is given on page 71 and window frames may be treated in a similar manner as shown in Figures 27(a) and (b). Window frames may be wedged at the top. Frames must be protected by priming, bedding and pointing as given for door frames.

Windows

The purpose of a window is to provide light and ventilation. The window should be so designed that the maximum amount of light is permitted to enter the building. Where ventilation is to be provided, which will be necessary in most habitable rooms, the Regulations require that an area equal to at least 1/20th of the floor area of the room must be able to be opened, at a height of not less than five feet nine inches above the floor. With casement windows it may not always be convenient to open the window at all periods of the year for ventilation, in which case a top hung fanlight will provide the necessary ventilation without penetration of rain, etc.

Timber Casement Windows

These are side hung windows which may be divided vertically into lights, or may be divided horizontally to provide a fanlight. The vertical members dividing the light are termed mullions and the horizontal member is termed a transome.

Sashes may be hung to open or may be fixed. A typical sash is shown in Figures 27(d) and (e), being exploded to show the joints between the members.

Sash or Glazing Bars

Where the area of glass is to be divided into smaller panes sash bars will be used. These should be of as small a section as possible so as not to restrict the entry of light. Figure 27(f) shows the intersection of two sash bars.

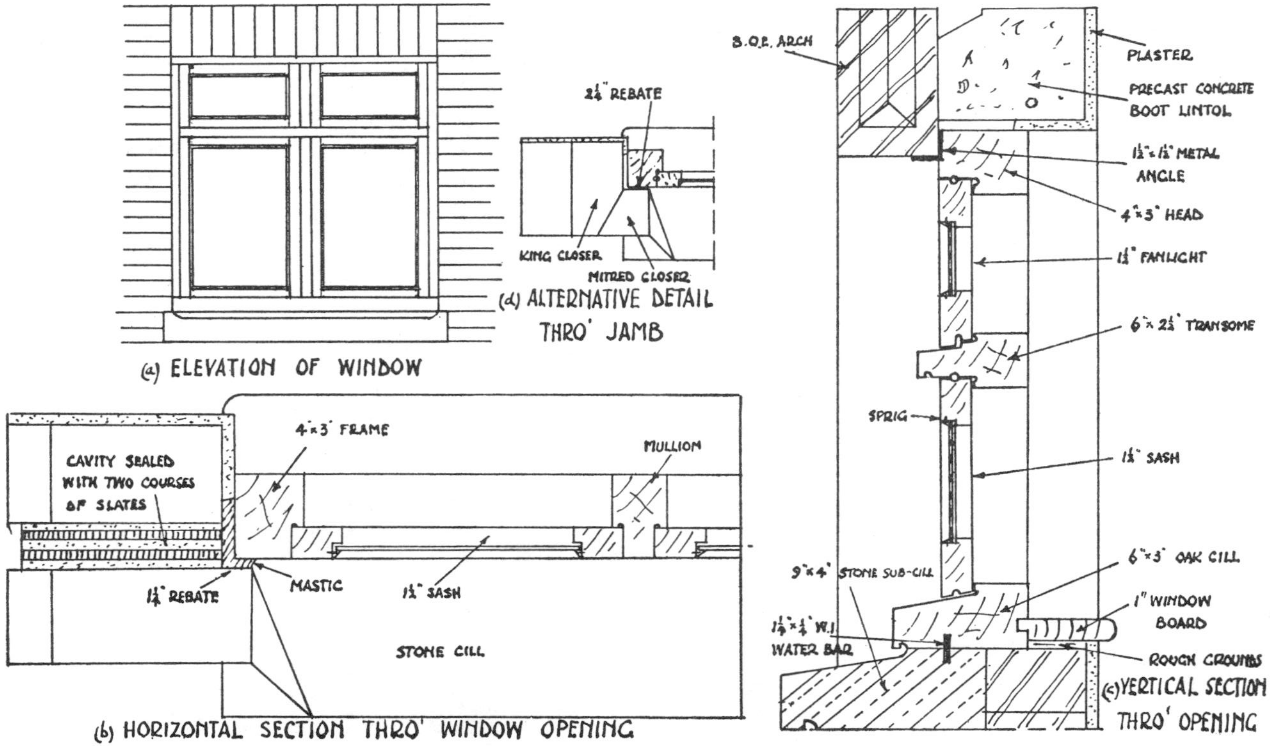
(a) ELEVATION OF WINDOW
2¼" REBATE
KING CLOSER
MITRED CLOSER
(d) ALTERNATIVE DETAIL THRO' JAMB
CAVITY SEALED WITH TWO COURSES OF SLATES
4"×3" FRAME
MULLION
1¼" REBATE
MASTIC
1½" SASH
STONE CILL
(b) HORIZONTAL SECTION THRO' WINDOW OPENING
S.O.B. ARCH
PLASTER
PRECAST CONCRETE BOOT LINTOL
1½"×1½" METAL ANGLE
4"×3" HEAD
1½" FANLIGHT
6"×2½" TRANSOME
SPRIG
1½" SASH
9"×4" STONE SUB-CILL
6"×3" OAK CILL
1" WINDOW BOARD
1¼"×¼" W.I. WATER BAR
ROUGH GROUNDS
(c) VERTICAL SECTION THRO' OPENING

Figure 28

Anti-capillary Grooves

Where exterior joinery is well made the surfaces of the timber which are in contact will induce capillary attraction, drawing water to the inside of the window. To overcome this anti-capillary grooves are provided in the frame and sash as shown in Figure 28.

Sill

As already stated on page 28, windows have a great proportion of glass, which is non-absorbent, causing the water to collect at the sill. The sill should be of hardwood and allowed to project beyond the brickwork, or supporting member, to throw the water away from the wall.

Transome

The transome is a combined sill and head, as it has to function as both, as seen in Figure 27(b), which includes the method of jointing to the frame.

Metal Windows

Domestic metal windows are normally of standard type based on B.S. 990 and can be obtained in widths of 11″, 1′ 8″, 3′ 3¼″, 4′ 10½″, and 6′ 6½″, and heights of 11½″, 2′ 0¾″, 4′ 0″, 4′ 11⅜″.

They may be fixed directly to brickwork or masonry, or fixed in a timber frame as shown in Figure 29, according to choice. The variety of standard windows can be seen in any manufacturer's catalogue.

Figure 29 shows a section through a typical metal window with casement and fanlight.

While metal windows can be of aluminium or bronze, steel is the most common. Steel metal windows in their early days had a bad name, because of rusting, but with the introduction of the hot dip process, rusting has been eliminated and broken window panes cannot be attributed to this cause.

Glazing

The method of glazing to external timber windows will be by painting the rebate of the sash, to prevent the oil from the putty being absorbed by the timber, after which a layer of putty will be placed in the rebate, which is termed 'back putty', forming a bed for the glass, The glass will be held in position by small headless tacks known as 'sprigs'. A splayed fillet of putty will be placed around the edge of the glass, to make it weather resistant as shown in Figure 27(c).

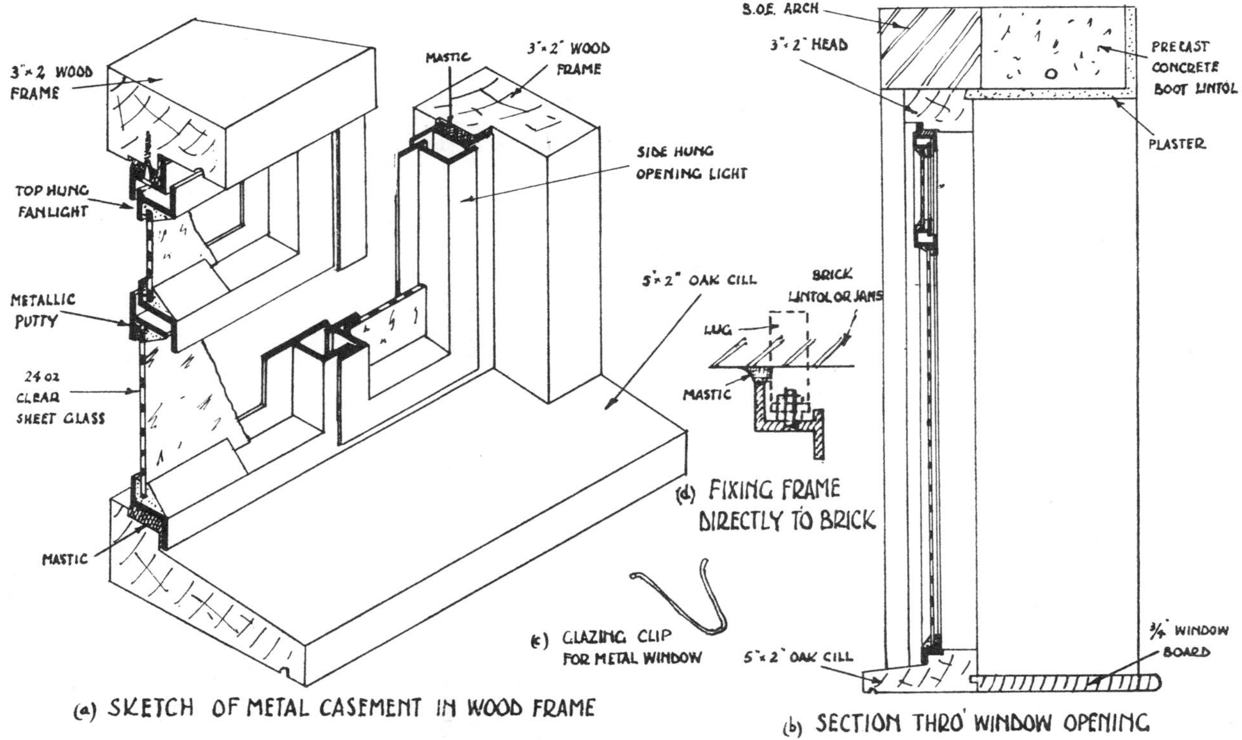
3" × 2 WOOD FRAME
TOP HUNG FANLIGHT
METALLIC PUTTY
24 oz CLEAR SHEET GLASS
MASTIC
MASTIC
3" × 2" WOOD FRAME
SIDE HUNG OPENING LIGHT
5" × 2" OAK CILL
(a) SKETCH OF METAL CASEMENT IN WOOD FRAME
BRICK LINTOL OR JAMS
LUG
MASTIC
(d) FIXING FRAME DIRECTLY TO BRICK
(c) GLAZING CLIP FOR METAL WINDOW
B.O.E. ARCH
3" × 2" HEAD
PRECAST CONCRETE BOOT LINTOL
PLASTER
3/4" WINDOW BOARD
5" × 2" OAK CILL
(b) SECTION THRO' WINDOW OPENING

Figure 29

Putty for timber casements will be linseed oil putty, while for metal windows metallic putty will be used, which is composed of red oxide.

Metal windows will be back puttied, but the glass will be held in by special clips, as shown in Figure 29(c), holes being provided in the casement for the clips. Putty splayed as before will make the edge of the glass weatherproof.

Glass

Manufacture

Glass is composed of sand (silica), soda and lime. These basic ingredients are mixed and melted to such a temperature that all the gases are given off and a clear liquid is obtained. From this point the process varies to provide the two main classes of glass available.

Sheet Glass

Used in buildings for windows, is produced by forcing the molten glass by hydrostatic pressure through a narrow slit in a slightly submerged float. A metal 'bait' is dipped in the liquid above the slit in the float and the glass is drawn up through metal cooling tanks into an annealing tower, where it is cooled and cut to size. This is known as the 'Fourcault Process'. The 'Libby-Owens' Process is similar to this but small rollers grip the edges of the sheet to maintain the width. A horizontal roller-lehr receives the glass when it enters the annealing tower and the glass is cut after passing over it in the later refinement of the above processes. This is known as the 'Modern Flat Drawn Process' and overcomes many of the faults occurring previously.

Flat drawn glass is manufactured in thicknesses from 18 oz to 32 oz per sq. ft, the variation in thicknesses being produced by the speed of drawing and the skill of the operator.

Plate Glass

The molten glass is drawn out in a continuous sheet between rollers, passed into a grinding machine, when both top and bottom surfaces are ground and afterwards polished. Thicknesses can range from $\frac{3}{8}''$ to $1\frac{1}{4}''$.

The use of plate glass in building work is usually in large windows or where safety is involved in such positions as doors, etc.

Rolled Glass

The molten glass is drawn out in a continuous sheet on to rollers, which convey it to an annealing lehr, after which it is cut.

To obtain the various types of obscure glass, patterns are applied with a patterned roller.

Wire glass is manufactured in a similar manner, wire mesh being introduced during the plastic stage when it leaves the molten condition in the tank.

Hanging Windows

Timber casement windows will be side hung on one pair of butts while fanlights will be top hung.

Fixed lights may be provided by fixing, instead of hanging, the sash, or the glass may be fixed to the frame without a sash.

Metal windows normally have easy clean hinges and are also side hung or top hung for fanlights.

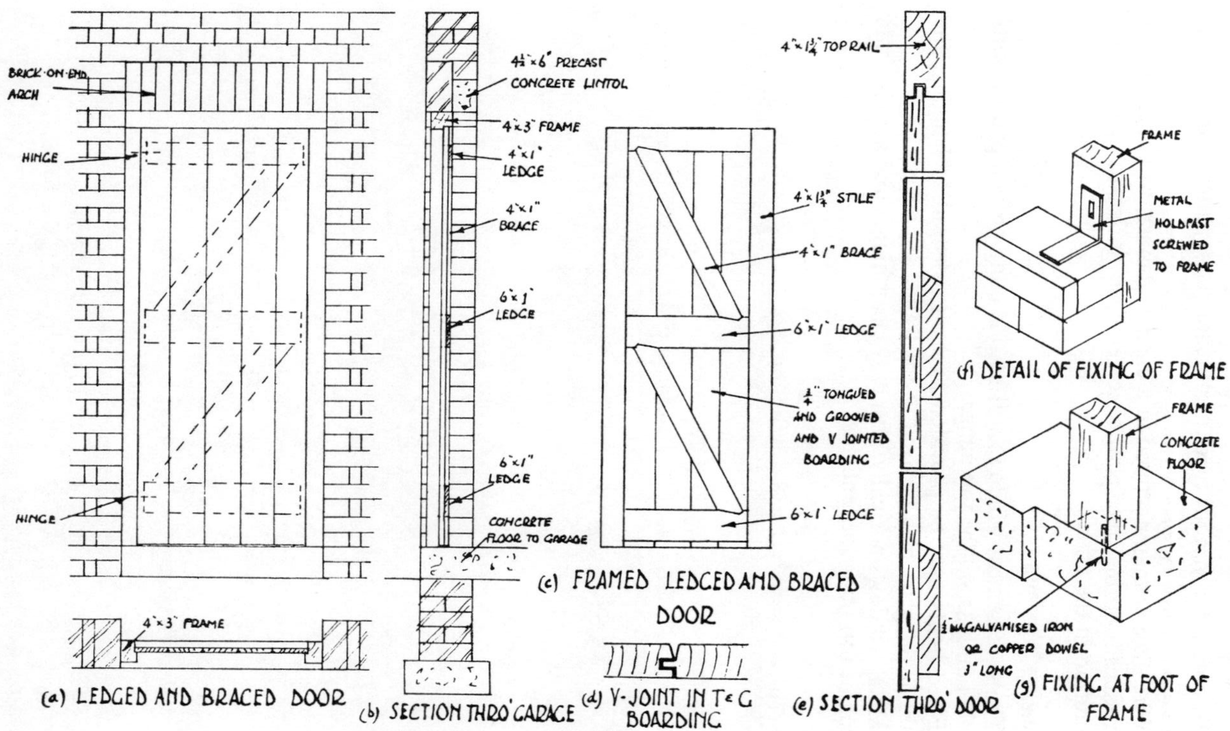
BRICK-ON-END ARCH
HINGE
HINGE
4"×3" FRAME
(a) LEDGED AND BRACED DOOR
4½"×6" PRECAST CONCRETE LINTOL
4"×3" FRAME
4"×1" LEDGE
4"×1" BRACE
6"×1" LEDGE
6"×1" LEDGE
CONCRETE FLOOR TO GARAGE
(b) SECTION THRO' GARAGE
4"×1¾" STILE
4"×1" BRACE
6"×1" LEDGE
¾" TONGUED AND GROOVED AND V JOINTED BOARDING
6"×1" LEDGE
(c) FRAMED LEDGED AND BRACED DOOR
(d) V-JOINT IN T&G BOARDING
4"×1¾" TOP RAIL
(e) SECTION THRO' DOOR
FRAME
METAL HOLDFAST SCREWED TO FRAME
(f) DETAIL OF FIXING OF FRAME
FRAME
CONCRETE FLOOR
½" DIA GALVANISED IRON OR COPPER DOWEL 3" LONG
(g) FIXING AT FOOT OF FRAME

Figure 30

11 Doors

Fixing Timber Frames

When the brick walling is built provision for fixing the frames should be made by building in two or three timber pads or breeze fixing bricks. The frame can be screwed or nailed back to these after the brickwork and plaster have dried. Breeze fixing bricks are normal sized bricks composed of lightweight aggregate, which will take nails. Should neither pads nor fixing bricks be provided the wall will require plugging, i.e. a hole is drilled into the brickwork and plugged with timber into which the fixing screw is driven. External door frames may also have a ½″ metal dowel 3″ long inserted in the bottom of the frame and bedded in the threshold, as shown in Figure 30(g).

In domestic work it is common to erect the frame before the brickwork to the sides of the opening is commenced. Two metal holdfasts are screwed to the frame for building into the brickwork, as shown in Figure 30(f). The disadvantage of this method is that the timber becomes saturated during the process of bricklaying and plastering, which can affect the timber.

Horns are provided on the frame to give strength to the joint and these can be built in or cut off before fixing.

All frames should be primed with red lead or boiled linseed oil before fixing, to protect the built-in portion of the frame, which cannot be painted later.

Frames are bedded in oil putty and pointed in mastic to prevent penetration of water. Mastic is red lead and linseed oil. There are also proprietary mastics, which can be applied with a gun.

Internal Linings

Where doors are to be provided on internal walls the opening will be lined, since the frame previously described would not adequately cover the wall behind. Figure 31 shows two methods of providing a lining to a 9″ wall, for a 4½″ wall the width would be reduced accordingly. In (*a*) the backings are dovetailed into the grounds, the lining is rebated for the door and the grounds covered

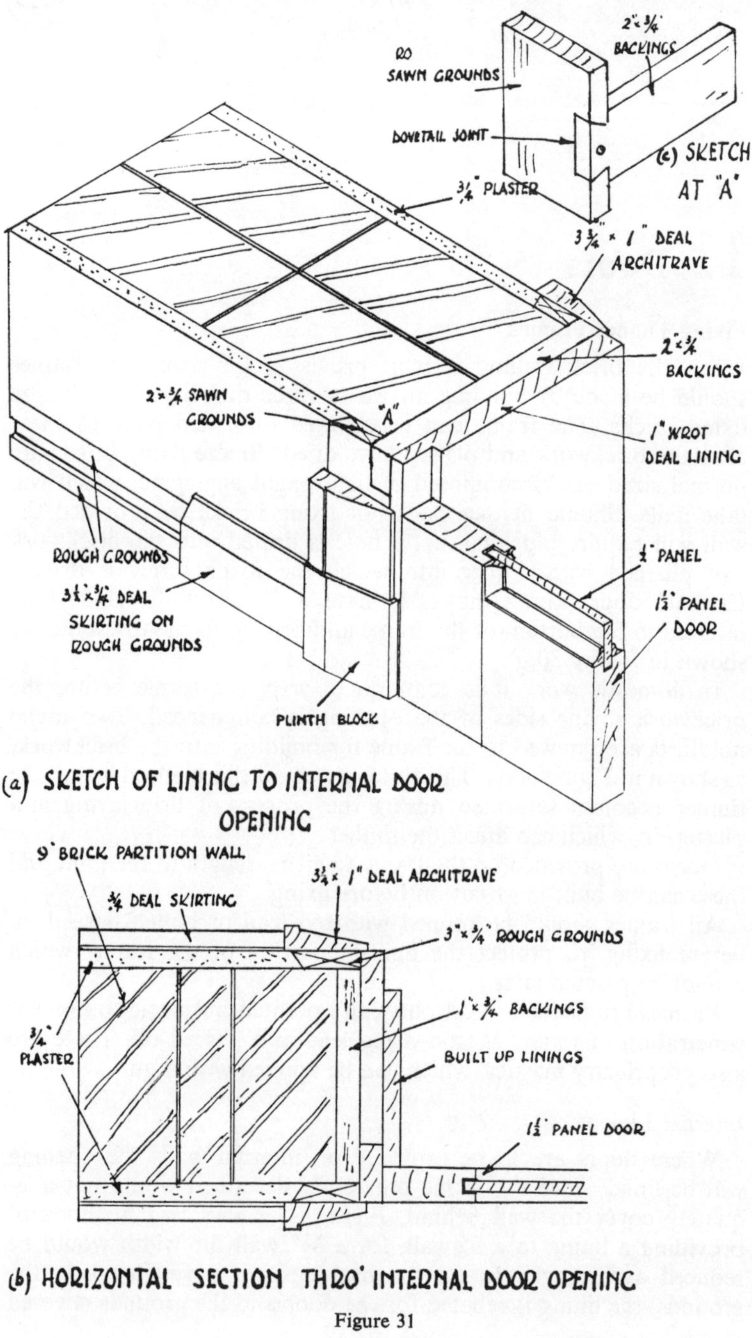

2" x 3/4" BACKINGS
RO SAWN GROUNDS
DOVETAIL JOINT
(c) SKETCH AT "A"
3/4" PLASTER
3 3/4" x 1" DEAL ARCHITRAVE
2" x 3/4" BACKINGS
2" x 3/4" SAWN GROUNDS
"A"
1" WROT DEAL LINING
ROUGH GROUNDS
1/2" PANEL
3 1/2" x 3/4" DEAL SKIRTING ON ROUGH GROUNDS
1 1/2" PANEL DOOR
PLINTH BLOCK
(a) SKETCH OF LINING TO INTERNAL DOOR OPENING
9" BRICK PARTITION WALL
3 3/4" x 1" DEAL ARCHITRAVE
3/4 DEAL SKIRTING
3" x 3/4" ROUGH GROUNDS
1" x 3/4" BACKINGS
3/4" PLASTER
BUILT UP LININGS
1 1/2" PANEL DOOR
(b) HORIZONTAL SECTION THRO' INTERNAL DOOR OPENING

Figure 31

by the architrave described on page 84. In (*b*) the lining is in three parts, being built-up to form the rebate for the door. The bonding of the reveals for a 9″ wall is also shown on the drawing.

Ledged and Braced Doors

The simplest type of door is the ledged door, which consists of a series of battens held together by two or three ledges. To resist the tendency for the door to drop on the side away from the hinges, a brace may be fixed between the ledges as shown in Figure 30(a). The direction of the braces must be as shown to be effective.

The strongest of these doors is the framed, ledged and braced door. This will consist of a frame, ledges, battens and braces. The door is of similar construction to a panelled door. A detail of this door is given in Figure 30(c).

The battens will be tongued and grooved and V-jointed. The accentuated joint will conceal any moisture movement of the battens.

Panel Doors

These may be from one panel upwards, each type having the same basic construction. Doors are generally of standard size in domestic work, being 2′ 3″ × 6′ 3″, 2′ 6″ × 6′ 6″, 2′ 8″ × 6′ 8″, 3′ 0″ × 7′ 0″.

Taking the four panel door as being the most common, it will consist of stiles, rails and muntins as shown in Figure 32(a). The stile will continue for the full height of the door and extend $1\frac{1}{2}$″ beyond at both top and bottom, these extensions of the stile are known as 'horns'. Their purpose is to allow additional material in making the joint at top and bottom between the rails and stiles, and also to protect the end of the stiles from damage during transport, storing and taking to position for fixing on site. The horns will be cut off before fixing.

The size of the members in a door will vary according to the architect's requirements. It is common for the stile, top rail and the muntin to be the same width, while the middle and bottom rails will be twice this width. The thickness of doors will be $1\frac{3}{4}$″ and 2″ for external use, and $1\frac{1}{2}$″ for internal use.

The joints between members and their names are shown in Figure 32. The haunch stiffens the joint and the twin haunched tenon allows a mortise lock to be used without weakening the joint.

The panels may be solid or of plywood, being let $\frac{3}{8}$″ deep into the stiles and rails. To cover the joint between the panel and the stile or rail a moulding should be provided. Several types of moulding

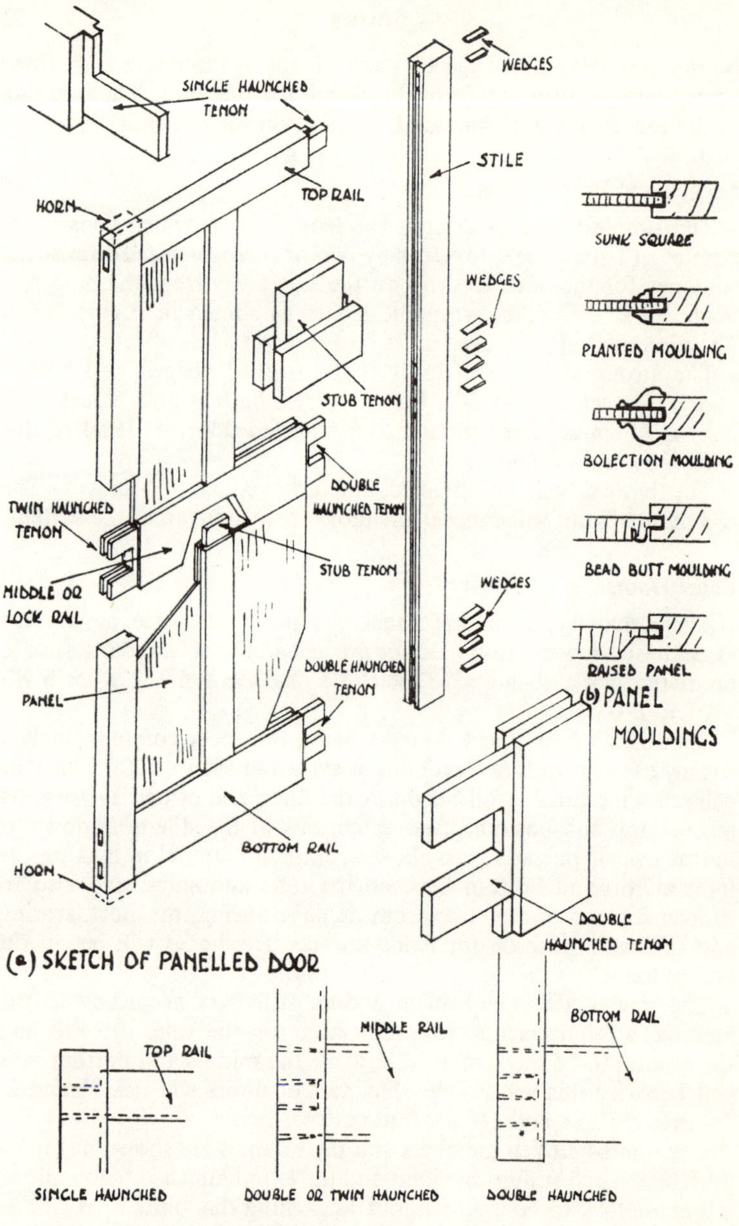
SINGLE HAUNCHED TENON
WEDGES
TOP RAIL
STILE
HORN
SUNK SQUARE
WEDGES
PLANTED MOULDING
STUB TENON
BOLECTION MOULDING
DOUBLE HAUNCHED TENON
TWIN HAUNCHED TENON
STUB TENON
BEAD BUTT MOULDING
MIDDLE OR LOCK RAIL
WEDGES
RAISED PANEL
DOUBLE HAUNCHED TENON
PANEL
(b) PANEL MOULDINGS
BOTTOM RAIL
HORN
DOUBLE HAUNCHED TENON
(a) SKETCH OF PANELLED DOOR
MIDDLE RAIL
BOTTOM RAIL
TOP RAIL
SINGLE HAUNCHED
DOUBLE OR TWIN HAUNCHED
DOUBLE HAUNCHED
(c) TENONS

Figure 32

are shown in Figure 32(b). Planted mouldings should be fixed to the stile or rail with panel pins and not to the panel, thereby hiding any movement of the panel after the door has been hung.

Flush Doors

The flush door has largely superseded the panel door in domestic construction. There are many proprietary makes on the market, using a variety of methods of manufacture. The basic construction of a flush door is a timber framework, upon which plywood is fixed to both faces.

The framing of the door will vary but a typical form is given in Figure 33. With the introduction of modern adhesives many of these doors are glued and not framed together, the plywood facing being tacked on and glued to the framing.

Many types of lightweight filling are also used, e.g. straw, wood shavings, etc.

The framing should provide for mortise locks and where flush doors are to be used in swing doors an inspection panel should also be incorporated.

Holes should be provided vertically through the framing for the height of the door to allow a free passage of air. This will overcome the twisting of doors, which sometimes takes place.

Ironmongery

There is a great variety of ironmongery available on the market, both as regards type and quality and the choice normally will be made by the Architect, based on the quality of the building or the position of the fitting. The following are some of the more common types of fitting suitable for the windows and doors given previously.

Butt Hinges (Figure 34(a)), upon which casement and fanlight windows as well as doors will be hung, are obtainable in a range of sizes and may be of steel, bronze, etc. A pair of butt hinges normally will be sufficient for a window or door, but the larger or heavier door will have a pair and a half.

Cross Garnet Hinges (Figure 34(b)), also known as Tee hinges, are suitable for ledged doors. Usually made of steel or japanned iron.

Strap Hinges. A stronger form of cross garnet, for use where the door is heavy and requires additional support, such as in garage doors.

Easy Clean Hinges. The usual fitting on metal casements, but can also be obtained for timber casements. The hinge has a long arm, which allows access to the outside of the casement for cleaning purposes.

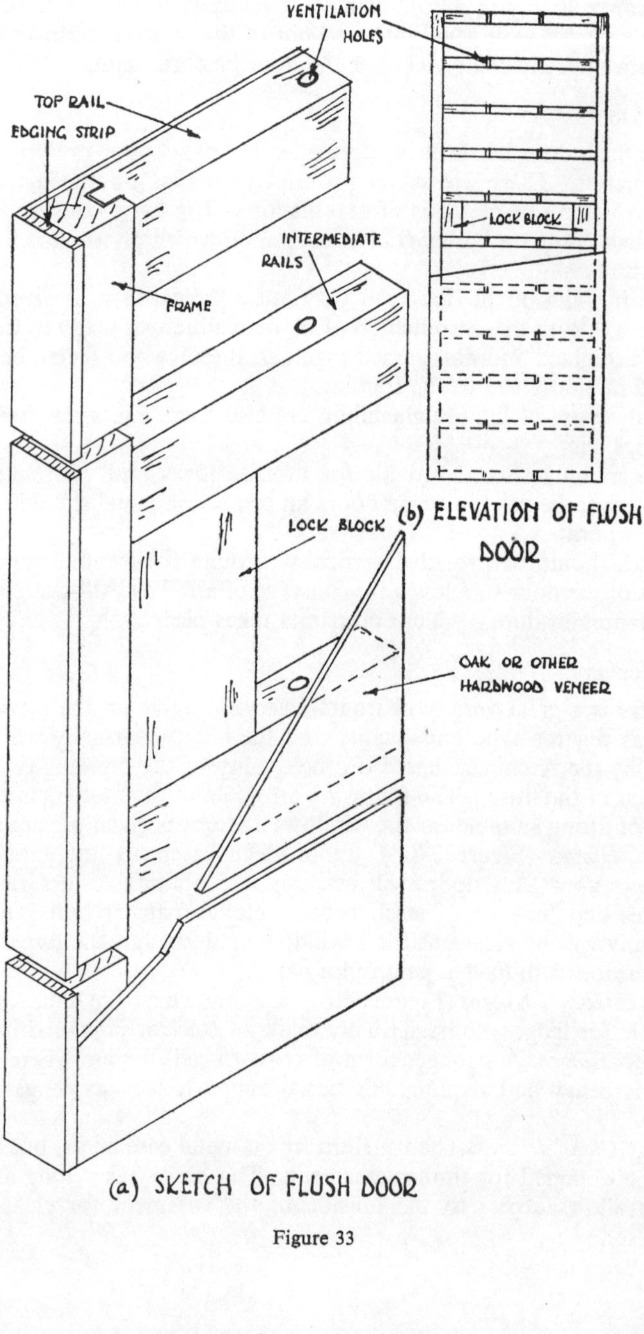
VENTILATION HOLES
TOP RAIL
EDGING STRIP
INTERMEDIATE RAILS
FRAME
LOCK BLOCK
OAK OR OTHER HARDWOOD VENEER
(a) SKETCH OF FLUSH DOOR
LOCK BLOCK
(b) ELEVATION OF FLUSH DOOR

Figure 33

Fanlight Stay and Pins. To support the top hung fanlight when open. A double pin secures the fanlight when closed. These will be similar to Figure 34(f). A fanlight catch is shown in Figure 34(c).

Quadrant Stay (Figure 34(h)). To support the fanlight when bottom hung. Will require a fanlight catch to secure the window when closed. The window may be operated by cords, pole or special opening gear.

Casement Fastener, Stay and Pin (Figures 34(d) and (f)). To secure the casement when closed. A casement stay and pin will secure the casement when open.

Mortise Deadlock (Figure 34(e)). A lock set in a mortise in the door having a key operated sneck. Often combined with a latch fitting to secure the door when not locked, as shown in Figure 34(g).

Rim Lock (Figure 34(k)). A key operated lock fitted on the face of the door, may also be fitted with a latch for holding the door when not locked.

Rim Latch. A latch fitted on the face of the door in a similar manner as given in Figure 34(k) for the rim lock, but operated with a handle and not a key.

Furniture (Figure 34(j)). The handle, spindle and plates necessary to open the door.

A large selection of materials and a variety of shapes are obtainable including wood, china, glass, plastic, metal.

Combination Night Latch (Figure 34(m)). Very common form of lock for external doors. Because of the variety of combinations there is a greater degree of security.

Letter Plates and Postal Knockers (Figure 34(p)). They should conform to B.S. 2911/1957 being not less than 8″ × 1$\frac{3}{4}$″ fitted not less than 2′ 6″ and not more than 4′ 9″ from the floor.

Finger Plates (Figure 34(n)). To protect the surface decoration of the door, being fixed above and below the lock.

Thumb Latches (Figure 34(q)). For doors to outhouses, gates, etc., where a lock is not required, a thumb latch will secure the door, but allow operation from both sides. There are a variety of these on the market.

Pull Handle (Figure (34(r)). For doors or cupboards not fitted with furniture.

Ball Catch (Figure 34(l)). To hold the door in a closed position, where a lock or latch is not desired. Fixed in a mortise in the stile of the door. A handle will be required on both sides of the door.

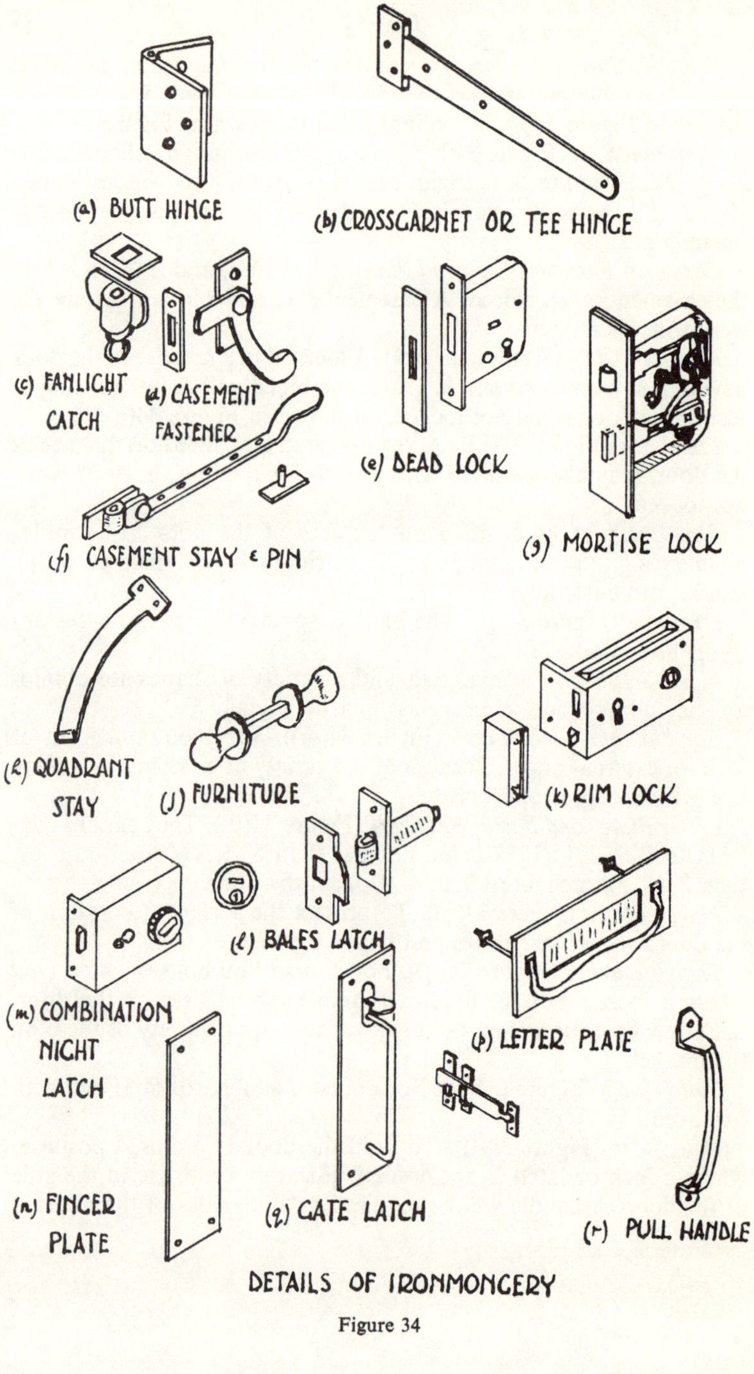
(a) BUTT HINGE
(b) CROSSGARNET OR TEE HINGE
(c) FANLIGHT CATCH
(d) CASEMENT FASTENER
(e) DEAD LOCK
(f) CASEMENT STAY & PIN
(g) MORTISE LOCK
(h) QUADRANT STAY
(j) FURNITURE
(k) RIM LOCK
(l) BALES LATCH
(m) COMBINATION NIGHT LATCH
(n) FINGER PLATE
(p) LETTER PLATE
(q) GATE LATCH
(r) PULL HANDLE
DETAILS OF IRONMONGERY

Figure 34

12 Finishings

Internal Finishings

The choice of finishings will be related to the use of the room. For most domestic work the covering to walls will be restricted to plaster and tiling, but in industrial and office buildings a large range of materials can be used.

Wall Finishings in Plaster

Lime plaster has been used greatly in the past for wall finishings, but has been largely superseded by gypsum wall plaster of recent years.

Lime plaster will be applied in three coats and will consist of plastering sand and lime in the proportions of one part of lime to four parts of sand for the backing coat, and one part of lime to two parts of sand for the finishing coats. See Figure 35(g).

Plastering sand must be thoroughly clean and free from impurities, particularly clay or salts, and should not be too sharp or too soft in texture.

Lime for plastering will be hydrated lime and is purchased in paper bags, being easily prepared and not requiring slaking. Lime plaster will have a backing, or rendering, coat $\frac{3}{8}''$ or $\frac{1}{2}''$ thick trowelled on to the brick wall. After completion the surface of the plaster will be scratched, or scored, with a scratcher consisting of a float with two or three protruding nails or with the point of the trowel, to form a key for the next coat. The second coat, known as the floating coat, will be $\frac{1}{4}''$ thick and will be trowelled on to the backing coat, after it has thoroughly dried and set. The surface of the floating coat will be scratched as previously. The third, or final coat, known as the finishing coat, will be $\frac{1}{8}''$ thick, trowelled to a smooth finish.

Gypsum plaster will consist of one of the several types of patent plaster on the market and with the addition of plastering sand will produce a hard wall plaster.

Gypsum is a rock consisting of calcium sulphate. It is ground and heated in a kiln which drives off most of the moisture. A hemi-hydrate is formed, which is known as Plaster of Paris. As this sets at varying times it is not suitable for general plastering work and a retarder is added producing a retarded hemi-hydrate plaster.

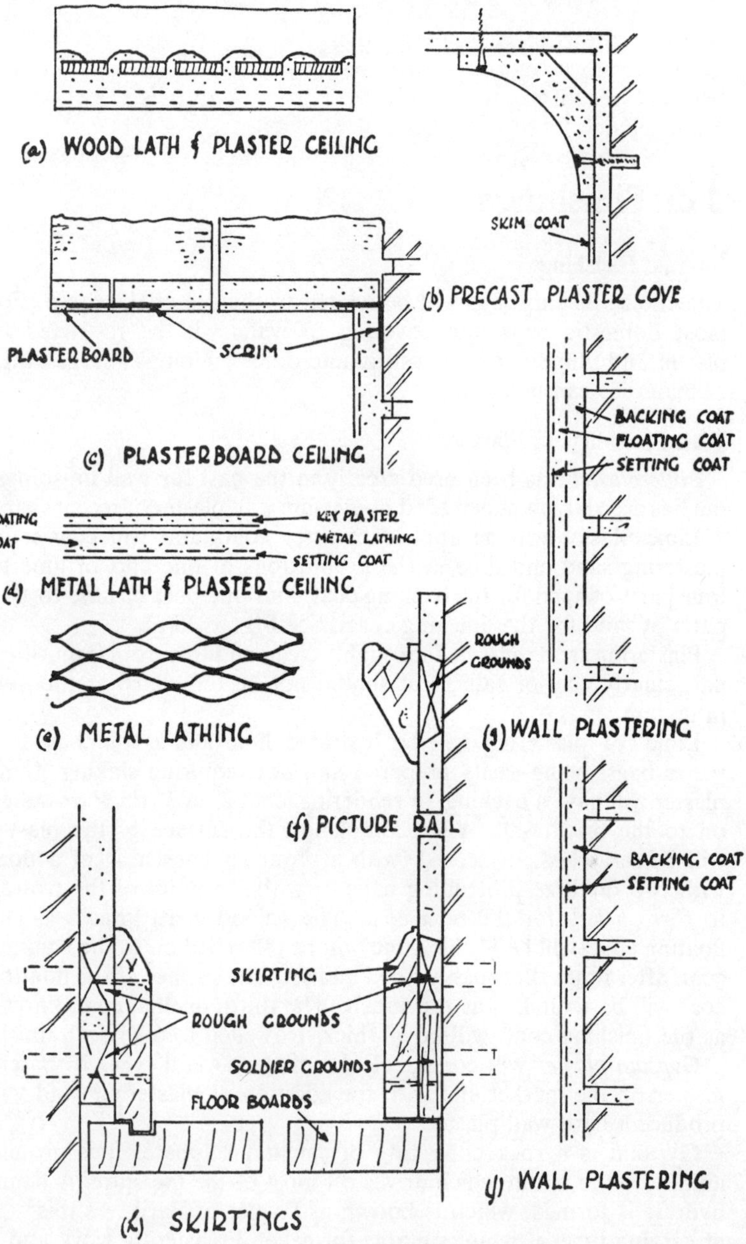
(a) WOOD LATH & PLASTER CEILING
SKIM COAT
(b) PRECAST PLASTER COVE
PLASTERBOARD
SCRIM
(c) PLASTERBOARD CEILING
BACKING COAT
FLOATING COAT
SETTING COAT
FLOATING COAT
KEY PLASTER
METAL LATHING
SETTING COAT
(d) METAL LATH & PLASTER CEILING
ROUGH GROUNDS
(e) METAL LATHING
(g) WALL PLASTERING
(f) PICTURE RAIL
BACKING COAT
SETTING COAT
SKIRTING
ROUGH GROUNDS
SOLDIER GROUNDS
FLOOR BOARDS
(j) WALL PLASTERING
(h) SKIRTINGS

Figure 35

An anhydrous plaster is a gypsum plaster which has had all the moisture removed and accelerators added to give a workable setting time. The resulting product is finely ground. Keenes cement and Parian cement are in this group.

An anhydrite plaster, of which Pioneer plaster is the most well known, is a similar plaster to the anhydrous, but takes longer to set.

Browning plasters are ready mixed plasters containing plaster and sand with only water being required to form the mix. Haired Browning plaster can also be obtained.

Gypsum plasters are applied in two coats, a backing, or rendering, coat $\frac{3}{8}''$ thick keyed to the wall, and a finishing coat $\frac{1}{8}''$ thick applied after the backing coat has set, the latter being scratched to provide a key for the final coat. See Figure 35(j).

Application of Plaster to Walls

To apply plaster to a wall surface it is necessary to provide guide points over the face of the wall to ensure that the final surface produced by the plaster is true.

To do this either:

(*a*) Dots are placed in plaster about 6″ below the ceiling level and the same distance from the floor, the whole being plumbed and made true for the final thickness. Other dots are placed at each angle of the wall and intermediate dots positioned. Vertical screeds, or bands of plaster, are then applied between the dots to the required thickness, the remainder of the wall is completed and made true to the screeds.

(*b*) Horizontal screeds are applied on the wall in bands at ceiling and floor positions and also at intermediate positions, according to the height of the wall. The screeds will be laid true and filled in between, after the screeds have set, to produce the wall surface. The finishing, or setting coat, is applied on top.

Ceilings

Laths. Where ceilings are of timber joist construction it will be necessary to provide a support and key for the plaster. The older form of this was the use of timber laths nailed with lath nails to the underside of the joists as shown in Figure 35(a). The width of the laths will be 1″ while the thickness varies according to the site requirements, and are classified as follows:

Single Laths	$\frac{1}{8}''$ thick
Lath and Half Laths	$\frac{1}{4}''$ thick
Double Laths	$\frac{3}{8}''$ thick

The laths can be either sawn or rift (or cleft) the latter being considered stronger because the fibres of the timber are not cut, but the lath is uneven in its outline.

Metal lathing, shown in Figure 35(e), is commonly used being fixed in sheets, the size of which will be 9′ 0″ × 2′ 0″ and is more adaptable than timber, particularly when the ceiling is curved.

Plastering to Ceilings

The plaster is applied in two or three coats according to whether patent plasters or lime plaster is used.

For lime plaster the first or backing or rendering coat $\frac{3}{8}$″–$\frac{1}{2}$″ thick is applied, the plaster penetrating the gap between the lathing and forming a key for the support of the whole ceiling, as shown in Figures 35(a) and (d). This coat is 'scratched' to form a key for the next coat and allowed to set. The next coat, that is the floating coat $\frac{1}{4}$″ thick, is applied, scratched and allowed to set. The final, or setting coat $\frac{1}{8}$″ thick is applied and finished to a true surface and allowed to set.

For patent plasters the above procedure applies but the floating coat is omitted.

Plasterboard

Most domestic houses will have plasterboard ceilings, because of the ease of fixing and the elimination of plaster mixes for the overhead work.

Plasterboard consists of precast gypsum plaster of the type given previously, formed in sheets 8′ 0″ × 2′ 0″ and compressed to give strength, and with one, or both sides, covered with thick paper.

The plasterboard is fixed to the underside of the timber joists with flat-headed nails and the joints between the sheets should be covered with hessian scrim held in position with plaster. See Figure 35(c).

Scrim should also be provided at the junction of the ceiling with the walls. The scrim will prevent cracking of the joints because of movement of the plasterboard under load from the joists above.

A skim coat of plaster $\frac{1}{8}$″ thick should be applied to the plasterboard to provide the finish.

Fibre Board

A softer ceiling finish can be provided by using fibre board. It is particularly useful where sound absorption and thermal insulation are required in a room and may also be used on the walls for this purpose. The fibre board sheets, from $\frac{1}{2}$″ thick, will be of size from 2′ 0″ wide and up to 18′ 0″ long and will be fixed to the timber joists in a similar manner as given for plasterboard, a $1\frac{1}{4}''$ panel pin being used. The joint between the boards can be covered with timber battens, or the edges of the boards splayed to give a V-joint. Battens or noggings can be fixed between the joists at the cross joints to provide a fixing for the boards, as shown in Figures 26(a) and (b). The surface of the fibre board can be distempered.

Variations of this type of board can be obtained to give various acoustic requirements.

Mouldings

Plasters used for plain internal work have been described, but additional decorations and finishes may be used in the form of mouldings.

Solid Mouldings

The application of a moulding at the junction of the wall and ceiling will improve the appearance of a room. These mouldings may be cast *in situ* or precast, the latter having many advantages. *In situ* solid mouldings will require greater skill to apply, particularly at the angles. Precast cove mouldings will be cast in the factory from gypsum plaster in convenient lengths up to 12 ft, and can be cut easily so that angles, whether internal or external, are no problem, since they can be mitred. For the larger room the lengths may be jointed and the joint made good with plaster. The cove will be plugged to the wall and screwed to the timber joists above, as shown in Figure 35(b).

Skirtings

A skirting will provide a finish between the floor and the wall and may be of timber, tile or other material matching the floor covering.

Timber skirtings should be fixed to grounds as shown in Figure 35(h). The grounds are of sawn timber plugged to the wall before plastering, i.e. a hole is drilled or punched into the wall, and a timber or plastic filling is driven into the hole, into which the fixing screw is located.

The purpose of the grounds which are termed '*first fixings*' are twofold. They will form a guide for the plasterer and also provide a fixing for the finished joinery work which is termed '*second fixings*'. By this method any dampness from the brickwork, or plasterwork, will only affect the sawn grounds. The joinery will not be fixed until after the grounds have dried out. Moisture will cause timber to move.

Skirtings to other floor finishings are shown in Figure 16.

Picture Rails and Dado Rails

These are not often incorporated in the modern domestic house, but should they be required, will be of wood and will be fixed as shown in Figure 35(f).

Architraves

Provide a finish to the grounds and frame, or lining, around a door or window. Timber architraves may be plain or moulded and will be fixed to the grounds and mitred at the angles, and may be provided with a plinth block at the foot, particularly when the skirting is to abut it. The plinth block will provide a stop for both members. See Figures 31(a) and (b).

External Finishings

The facing brick provides the usual form of external finish for the domestic house but other finishes can be used.

Cement Rendering

This is termed 'Plain face' and will consist of two coats of rendering, the backing coat $\frac{3}{8}''$ thick of one part cement and three parts sand and the finishing coat $\frac{1}{8}''$ thick of one part cement and $2\frac{1}{2}$ parts sand trowelled to a smooth face. The brickwork to which the rendering is to be applied will be left rough or built with keyed bricks. The problem of this type of finish is the tendency for it to craze and disfigure after completion, and it has now been found that a rough surface, which will hide any crazing, gives a more satisfactory finish. Examples of this rougher texture are:

(1) *Pebble dash finish*. The backing coat is applied and before it has set selected pebbles, shingle or spa is thrown on to the surface to give it a rough texture. The pebbles or gravel may be left or may be coated with a cement paint.
(2) *Rough cast*. The cement rendering is brought to an even finish and is afterwards roughened up with the trowel to give a texture to the face.
(3) *Machine finish rough cast*. The application of a rendering by a machine, which throws the finish on to the wall. This is also known as 'Tyrolean' finish.

13 Decoration

Distemper and Emulsion Paints

Ceiling and Walls

The application of distemper to these surfaces provides a clean finish in a variety of colours. Distemper can be washable and non-washable.

Washable distempers consist of barytes which has an oil-bound base, while non-washable distempers consist of whiting with a water base.

The introduction of emulsion paints has superseded the use of distemper for many classes of work. These consist of a plastic base.

Walls should be sized, i.e. covered with a layer of weak glue, before distempering, but size should not be used for emulsion paints.

Painting

Externally

Paint is used not only as a decoration, but as a protection of the material upon which it is applied. Timber will absorb moisture in its natural state and will shrink and swell according to this moisture content. In consequence the application of paint will protect the timber against moisture penetration. Where the timber surface is not seen as in frames to doors or windows, a coating of boiled linseed oil, as already stated on page 71, will protect the back. For surfaces which are visible the timber should be prepared by cleaning off the timber face and applying a coat of shellac, known as 'knotting', to any knots which may be in the timber.

If knotting is not applied the resin in the knot may force its way through the paint work. Any blemishes, nail holes, etc., should be filled with linseed oil putty or proprietary stopping to make an even surface to receive the paint. It should be remembered that paint will not cover blemishes but will tend to accentuate them. A priming coat of red lead, followed by two undercoats and one finishing coat of paint may be applied. Each undercoat should be rubbed down with fine glasspaper.

For ironwork the surface should be cleaned of all rust and grease and, if necessary, coated with anti-rust paint, followed by one or two undercoats and a finishing coat of paint.

Most good quality paint will have a lead base and will be composed of the following:

Base—ground lead to give body to the paint.
Vehicle—linseed oil to make it workable.
Pigment—animal or vegetable colouring.
Drier—to control the speed of drying of the paint.

The high quality of ready made paint available makes it unnecessary for the painter to prepare his own paint.

Internally

Paint applied internally is primarily used for decoration, its durability being related to wear and tear during normal domestic use. The application of the paint will be similar to that given for external use. The number of coats may be reduced to two by omitting one undercoat. The undercoat is important to give the depth, or body, of covering of the surface, the finishing coat providing in many cases the gloss required without great depth of covering, this particularly applies to white gloss paint.

Paint should be applied over the surface with a cross movement of the brush to remove brush marks and give a smooth finish to the member.

Varnish

Provides a coat of clear hard resin copal with oil and driers. There are two qualities, external, which provides a hard finish and internal, which is not so durable, but is glossy, hard and quick drying.

Varnish is added to paint to produce enamels.

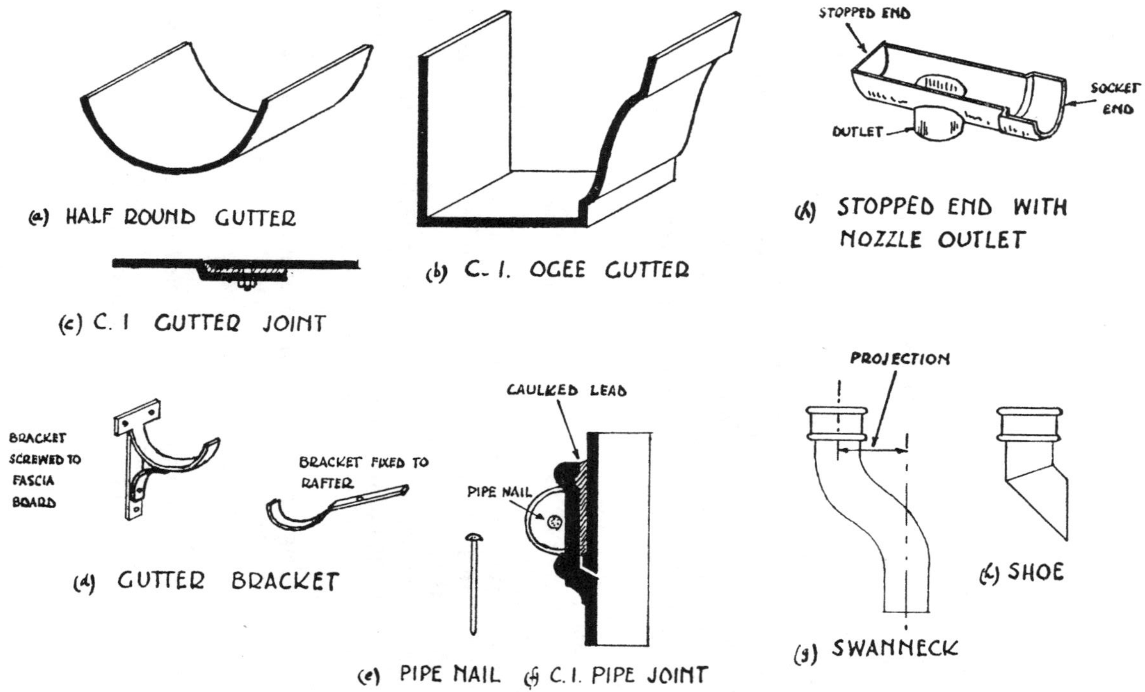
(a) HALF ROUND GUTTER
(b) C. I. OGEE GUTTER
STOPPED END
SOCKET END
OUTLET
(h) STOPPED END WITH NOZZLE OUTLET
(c) C. I GUTTER JOINT
PROJECTION
CAULKED LEAD
BRACKET SCREWED TO FASCIA BOARD
BRACKET FIXED TO RAFTER
PIPE NAIL
(d) GUTTER BRACKET
(k) SHOE
(g) SWANNECK
(e) PIPE NAIL
(f) C.I. PIPE JOINT
Figure 36

14 Plumbing

External Plumbing

Rainwater Removal

It is important that rainwater should be removed from the surface of a building as quickly as possible, and not allowed to settle or 'pool', which may encourage penetration to the inside, where damage could occur without it being observed until it was too late. Roofs and other sloping surfaces should be adequately pitched for the material used and having provided for this, water must be removed from this position of collection as speedily as possible.

Eaves Gutter

The first position of collection from roofs will be the gutter, which in the case of pitched roofs will be of cast iron, asbestos-cement or plastic and may be half round or ogee section. Details of various finishes at eaves are given in Figure 20.

Eaves gutters are fixed in several ways, the most common being by brackets, as shown in Figure 36(d). Brackets can vary considerably, from ornate types to the common type given. The gutter will be fixed to a slight fall towards the outlet and the fall will be obtained when the gutter brackets are fixed to the fascia board.

Gutters will be of spigot and faucet type as shown in Figure 36(a), (b) and (c), the spigot being placed in the faucet and bedded in mastic and fixed together by a gutter bolt. The mastic in the case of asbestos-cement will be a special asbestos compound, and in the case of plastic will be a filler.

Care must be taken with asbestos-cement that the gutter is not damaged by placing ladders against it or otherwise knocking it.

The fittings required for a gutter are angles, stopends and outlets, all as shown in Figure 36(h).

Rainwater Pipe

This is referred to as a 'fall pipe' or 'down pipe' in certain districts, and will be of cast iron, asbestos-cement or plastic. The same care must be exercised with asbestos-cement as mentioned previously.

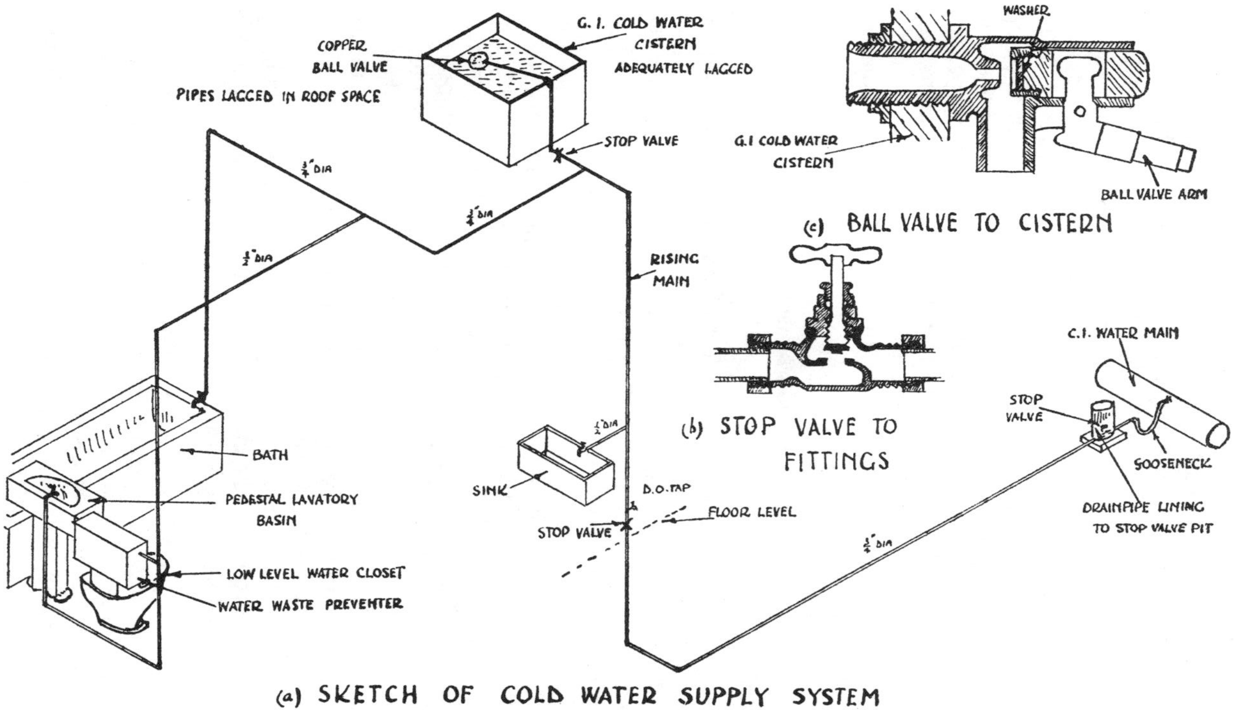
G. I. COLD WATER CISTERN ADEQUATELY LAGGED
COPPER BALL VALVE
PIPES LAGGED IN ROOF SPACE
STOP VALVE
¾" DIA
¾" DIA
½" DIA
RISING MAIN
BATH
PEDESTAL LAVATORY BASIN
LOW LEVEL WATER CLOSET
WATER WASTE PREVENTER
SINK
½" DIA
D.O. TAP
STOP VALVE
FLOOR LEVEL
¾" DIA
C.I. WATER MAIN
STOP VALVE
GOOSENECK
DRAINPIPE LINING TO STOP VALVE PIT
WASHER
G.I COLD WATER CISTERN
BALL VALVE ARM
(c) BALL VALVE TO CISTERN
(b) STOP VALVE TO FITTINGS
(a) SKETCH OF COLD WATER SUPPLY SYSTEM

Figure 37

The passage of water from the eaves gutter to the pipe will be by a swan-neck as shown in Figure 36(g), and the discharge into the gulley at the foot of the pipe will be by a shoe. See Figure 36(h).

Rainwater pipes will be of the spigot and socket type, with the socket uppermost to receive the spigot, the joint will be made of red lead in the case of cast iron, asbestos compound in the case of asbestos-cement, and a filler in the case of plastic. See Figure 36(f).

Other fittings may be required for rainwater pipes, such as plinth bends, where the pipe is to project for a plinth or other obstruction in the vertical face of the building. Rainwater heads, the type, size and variety of which are considerable, are used where more than one pipe is to discharge into a rainwater pipe. Often this is useful when a bath or lavatory basin is on the upper floor and is adjacent to a rainwater pipe.

Internal Plumbing

Cold Water Supply

Every building will require a cold water supply and this will be taken from the water main in the roadway, in the case of town or city. In country areas the supply of water will depend upon the degree of isolation of the building from built-up areas. A well is often used in such situations.

The cast iron water main in the roadway will be tapped and this will be carried out without dislocating the supply in the area.

The water supply to the building may be of lead, copper or plastic pipe. It is advisable that a gooseneck be provided, as shown in Figure 37(a), to allow for any tension and movement in the pipe when back filling of excavated material takes place. The trench should be sufficiently deep to be unaffected by the severest frost and the pipe should enter the building below floor level and rise within the building, a stop valve being provided at the lowest position above ground floor level. See Figure 37.

The layout of the cold water supply system within the building will vary according to the custom of the area where the building is situated and this will be in two forms:

(*a*) The supply will be taken directly to a storage tank in the roof space and all domestic fittings, with the exception of the drinking water, which must be taken from the rising main, will be supplied by distribution pipes from the tank.

(*b*) The supply will be taken to the highest located fitting and all other fittings will be supplied directly from the rising main, as shown in Figure 37.

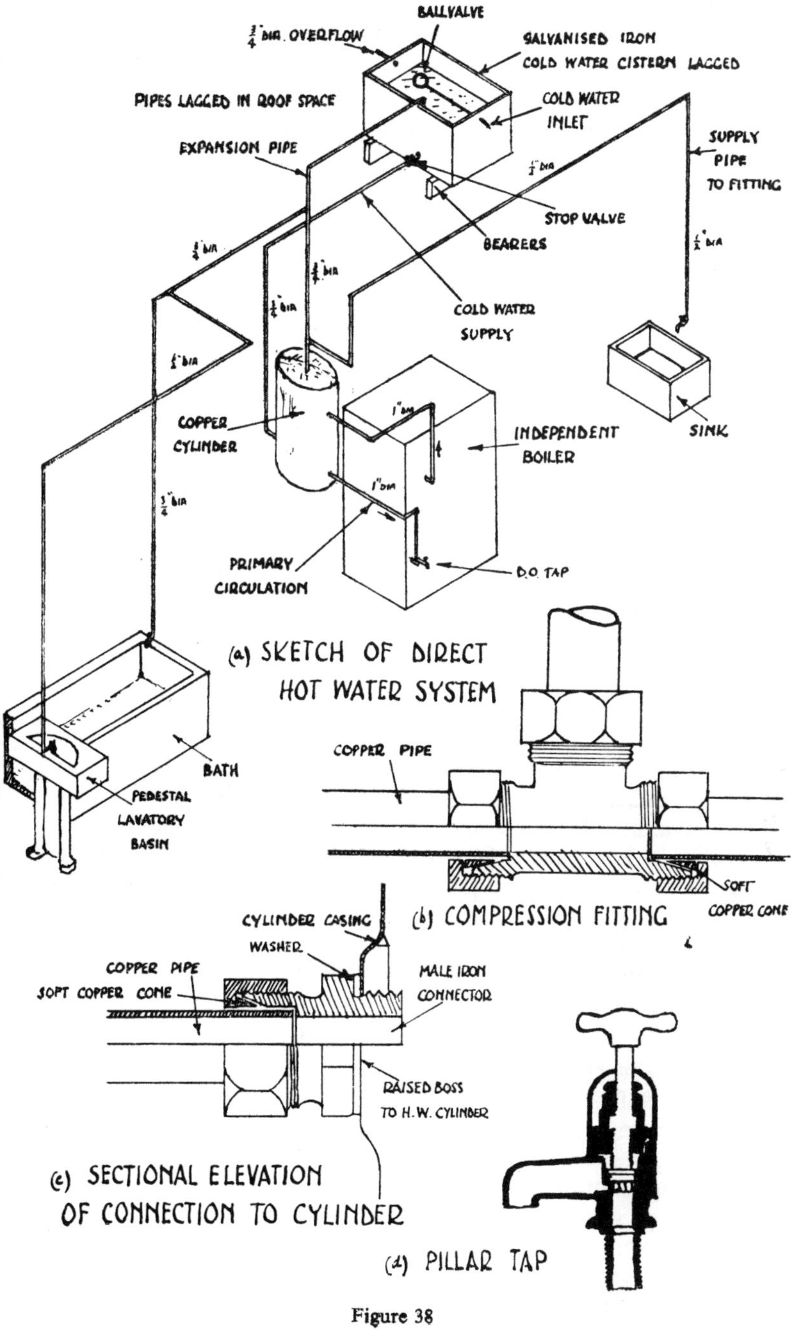
BALLVALVE
3/4" DIA. OVERFLOW
GALVANISED IRON
COLD WATER CISTERN LAGGED
PIPES LAGGED IN ROOF SPACE
COLD WATER
INLET
SUPPLY
PIPE
TO FITTING
EXPANSION PIPE
1/2" DIA
STOP VALVE
BEARERS
3/4" DIA
3/4" DIA
1/2" DIA
3/4" DIA
COLD WATER
SUPPLY
1/2" DIA
COPPER
CYLINDER
1" DIA
INDEPENDENT
BOILER
SINK
3/4" DIA
1" DIA
PRIMARY
CIRCULATION
D.O. TAP
(a) SKETCH OF DIRECT
HOT WATER SYSTEM
BATH
PEDESTAL
LAVATORY
BASIN
COPPER PIPE
SOFT
COPPER CONE
(b) COMPRESSION FITTING
CYLINDER CASING
WASHER
COPPER PIPE
SOFT COPPER CONE
MALE IRON
CONNECTOR
RAISED BOSS
TO H. W. CYLINDER
(c) SECTIONAL ELEVATION
OF CONNECTION TO CYLINDER
(d) PILLAR TAP

Figure 38

The internal water pipe will be of lead or copper $\frac{1}{2}''$ diameter, the fittings for copper will generally be of the compression type, rather than screwed fittings, as shown in Figure 38(b).

Hot Water System

Hot water can be provided domestically in one of several ways, the simplest form being by instantaneous supply from a gas or electric geyser, where the water is heated, as required, by passing over a hot surface. This is suitable for sinks and baths, etc.

Where constant hot water is required at several points in the building a storage cylinder is necessary. The following gives the most common form of heating for this type of supply.

Direct Supply

The water is heated by a boiler, which can be fired by solid fuel, gas, or oil, and the heated water is stored in a storage cylinder. The various fittings are supplied from the storage cylinder as shown in Figure 38. The boiler which can be used in this system, can be either a back boiler or an independent boiler, according to the requirements of the building and the capacity required. An electricity immersion heater can also be used instead of a boiler.

Pipes

The use of lead in pipes for domestic hot water has been superseded by copper, because of cost. Care must be exercised in the use of different metals in combination, because of the possibility of electrolysis taking place. As already stated, for cold water services copper tube may be screw jointed or capillary joints may be used. The use of plastics for hot water circuits has certain difficulties and is not generally used.

Lagging of Pipes

Pipes which are exposed and subject to attack by frost should be protected by lagging and any pipes or water tanks in the roof space should be similarly protected. With hot water circuits, particularly where a heating system is incorporated, any pipes which are positioned to give large heat losses, without adding to the warmth of the building, should be lagged. Hot water pipes in such positions as under the ground floor or in the roof space would be more efficient if lagged. The lagging should be of fibre glass or other similar non-conductive material which can be bought in sections and clamped on to the pipe or wrapped around it.

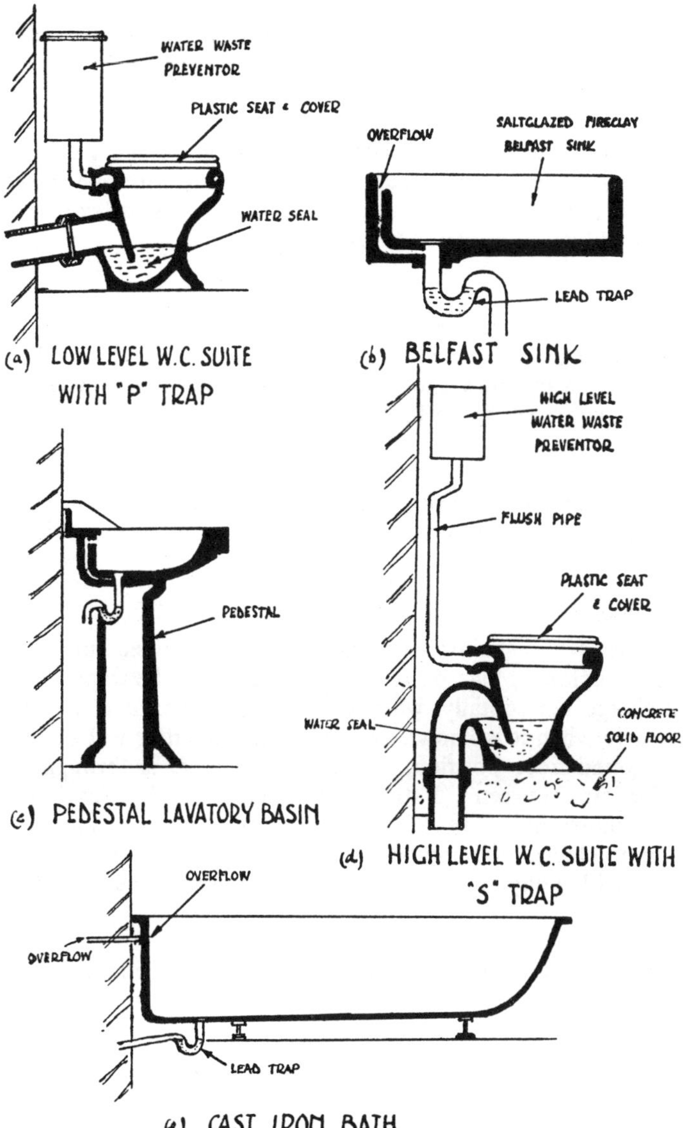
WATER WASTE PREVENTOR
PLASTIC SEAT & COVER
WATER SEAL
(a) LOW LEVEL W.C. SUITE WITH "P" TRAP
OVERFLOW
SALTGLAZED FIRECLAY BELFAST SINK
LEAD TRAP
(b) BELFAST SINK
PEDESTAL
(c) PEDESTAL LAVATORY BASIN
HIGH LEVEL WATER WASTE PREVENTOR
FLUSH PIPE
PLASTIC SEAT & COVER
WATER SEAL
CONCRETE SOLID FLOOR
(d) HIGH LEVEL W.C. SUITE WITH "S" TRAP
OVERFLOW
OVERFLOW
LEAD TRAP
(e) CAST IRON BATH

Figure 39

Domestic Sanitary Fittings

The usual fittings found in the domestic house will be the sink, bath, lavatory basin and water closet.

Sink

This can be of white glazed clayware of one of a number of sizes, but in many modern houses the stainless steel sink with one or two bowls, and the vitreous enamel sink are now becoming common. See Figure 39(b) for a section through a sink.

Bath

A great variety of designs of baths can be obtained but they will all be of cast iron with or without a panel finish. The outlet will be trapped and an overflow should be provided. See Figure 39(e) for section through one type of bath.

Lavatory Basin

Can be obtained in various sizes to suit the position. Will be of white or coloured glazed fireclay. A section through a typical basin is shown in Figure 39(c).

Water Closet

A number of designs for water closets can be obtained but will be of white or coloured glazed fireclay. The outlet will be trapped and can discharge horizontally, when the trap is known as a 'P' trap, or vertically when it is known as a 'S' trap. The latter will be used in the ground floor position, where the discharge is through the floor. The joint between a 'P' trap and soil pipe for a low level suite will be as shown in Figure 39(a). An alternative high level suite with a 'S' trap to clayware drainpipe is shown in Figures 39(d) and 40(g). The latter with a high or low level suite will be used in the bungalow.

Plumbing Joints

Figure 40 shows joints between various types of pipe in plumbing work.

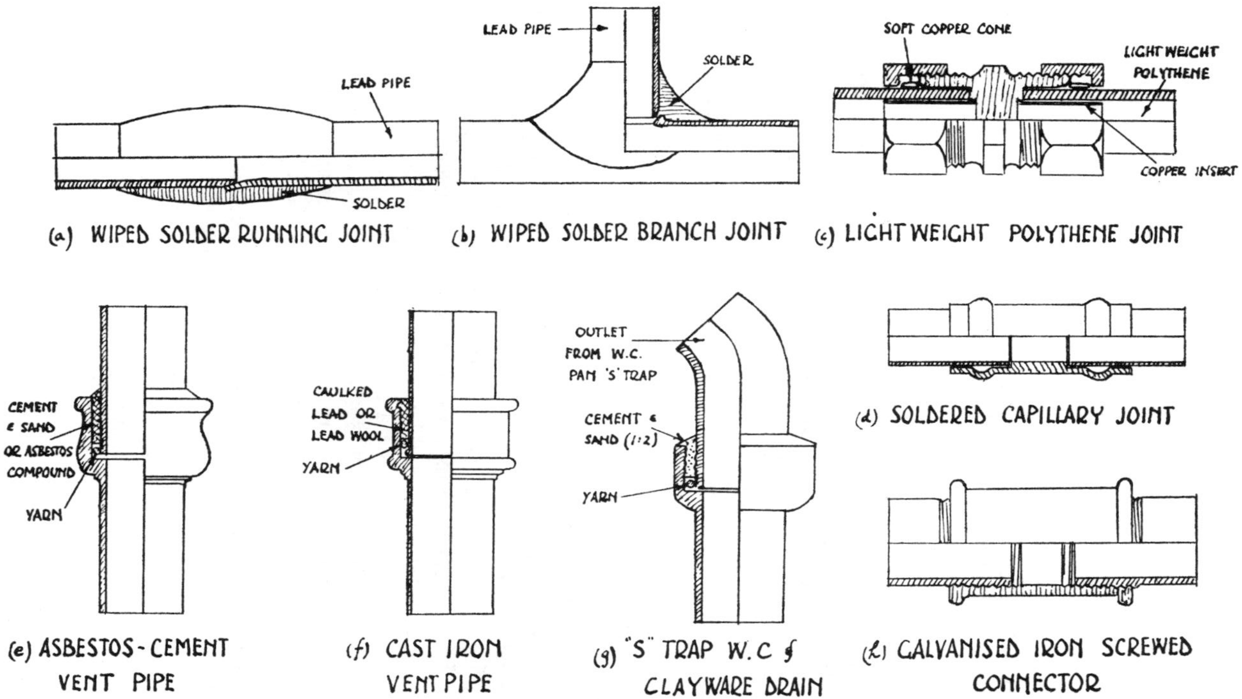
LEAD PIPE
SOLDER
(a) WIPED SOLDER RUNNING JOINT
LEAD PIPE
SOLDER
(b) WIPED SOLDER BRANCH JOINT
SOFT COPPER CONE
LIGHT WEIGHT POLYTHENE
COPPER INSERT
(c) LIGHT WEIGHT POLYTHENE JOINT
CEMENT & SAND OR ASBESTOS COMPOUND
YARN
(e) ASBESTOS-CEMENT VENT PIPE
CAULKED LEAD OR LEAD WOOL
YARN
(f) CAST IRON VENT PIPE
OUTLET FROM W.C. PAN 'S' TRAP
CEMENT & SAND (1:2)
YARN
(g) "S" TRAP W.C & CLAYWARE DRAIN
(d) SOLDERED CAPILLARY JOINT
(h) GALVANISED IRON SCREWED CONNECTOR

Figure 40

15 Drainage Systems

Rainwater Disposal

THE collection of rainwater from the building has already been dealt with, its eventual disposal will be governed by the requirements of the local authority. Some authorities require rainwater to be collected separately from foul water, and discharged into local brooks, streams, etc. Other authorities make provision in their sewage disposal for the disposal of rainwater. According, therefore, to these requirements one or two systems of drainage will be necessary.

All rainwater will discharge into an open gulley and not directly into the drainage system.

Foul Water Disposal

Waste water from the sink, lavatory basin, bath, urinal and water closet will discharge into the foul water drain and not into the storm-water drain given previously. The discharge from sink, lavatory basin and bath will be into an open gulley, but discharge from the urinal and water closet will be directly into the drain.

It is advisable to provide a trap on all fittings, whether required by the Regulations or not, some fittings, discharging into an open gulley, will collect odours from the gulley at certain times of the year and transfer these to the inside of the building. The gulley into which the fittings will discharge will have a trap, thus safeguarding the building from any sewer gases which may be present.

The urinal and water closet which discharge directly into the drain, will be provided with a trap as shown in Figure 42(c), the former in the outlet and the latter in the pan and these will be the only seals from the drain. Care must be taken that these seals are not broken, particularly in the case of the pan.

The foul water will be taken from the building by pipes to a manhole. It will be usual to carry this waste in saltglazed stoneware pipes of 4″ diameter, jointed with cement and sand as shown in Figure 42(a).

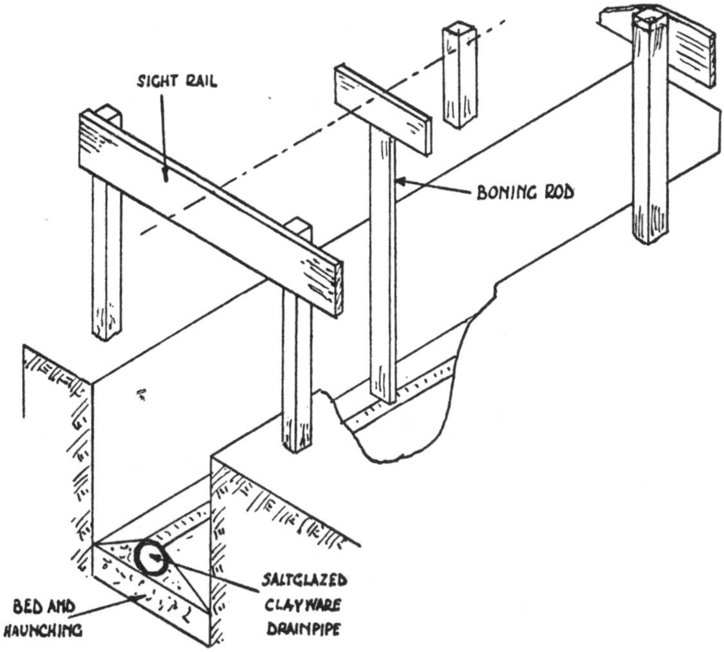

(a) SIGHT RAILS FOR DRAIN TRENCH

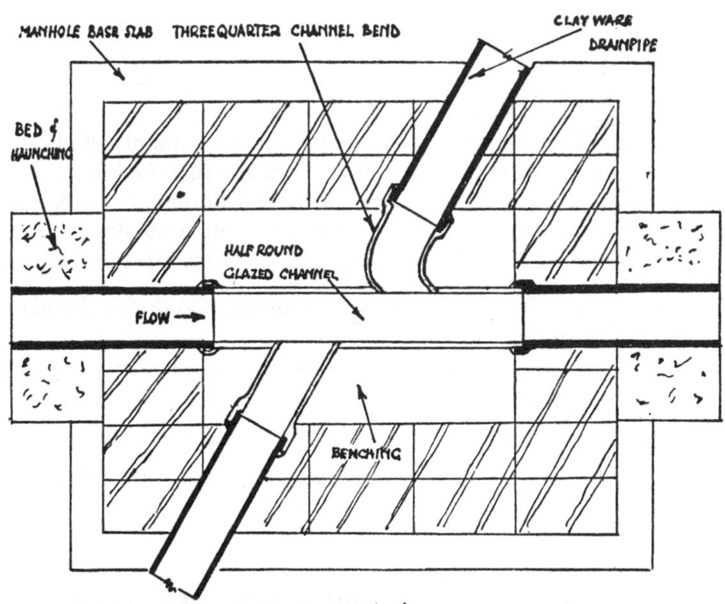

(b) HORIZONTAL SECTION THRO' MANHOLE

Figure 41

The design of a drainage system should be such that the lengths of drain are in straight lines for easy rodding. Short branches to fittings should be steep to keep them flushed. The gradient of main runs of pipes should be sufficient to allow the water and solids to clear quickly but being not too steep that the solids are left behind.

Where branches meet the main line of drain, junctions are provided to connect the two pipe runs.

Trenches for Drains

These will be excavated by hand or by machine according to the lengths of drain to be provided.

Drainpipes will be laid to a fall, which will vary according to their diameter. The usual fall for drainpipes will be ten times their nominal diameter, thus for a 4″ diameter drainpipe the fall should be 1:40.

Trench bottoms will, therefore, be required to fall to suit the pipe requirements. To provide for this in excavating the trench sight rails and boning rods are used. Sight rails consist of a board supported at each end on posts and fixed across and above the trench. Several boards and posts are fixed at intervals along the trench, the top of each board being 'lined in' to suit the fall. As the trench is excavated the bottom is measured from the top of the board by boning rods to obtain a regular fall. These rails and boards are shown in Figure 41(a).

Drainpipes

Saltglazed clayware drainpipes are the most commonly used both for storm and foul drains; pitch fibre pipes are also used for storm-water drains. Cast iron pipes are used when drains are to be laid under buildings or where the pipes may be liable to damage.

There are several qualities of saltglazed clayware pipes as follows:

1. Individually Air tested pipes conforming to the British Standard Specification.
2. Ditto, but not specified as conforming to the British Standard.
3. British Standard Tested pipes conforming to the British Standard Specification, each pipe having passed the hydraulic test.
4. British Standard pipes conforming to the British Standard Specification, a proportion of these pipes being tested if requested.
5. Tested pipes. Every pipe has passed the hydraulic test but does not claim to conform to the British Standard Specification.

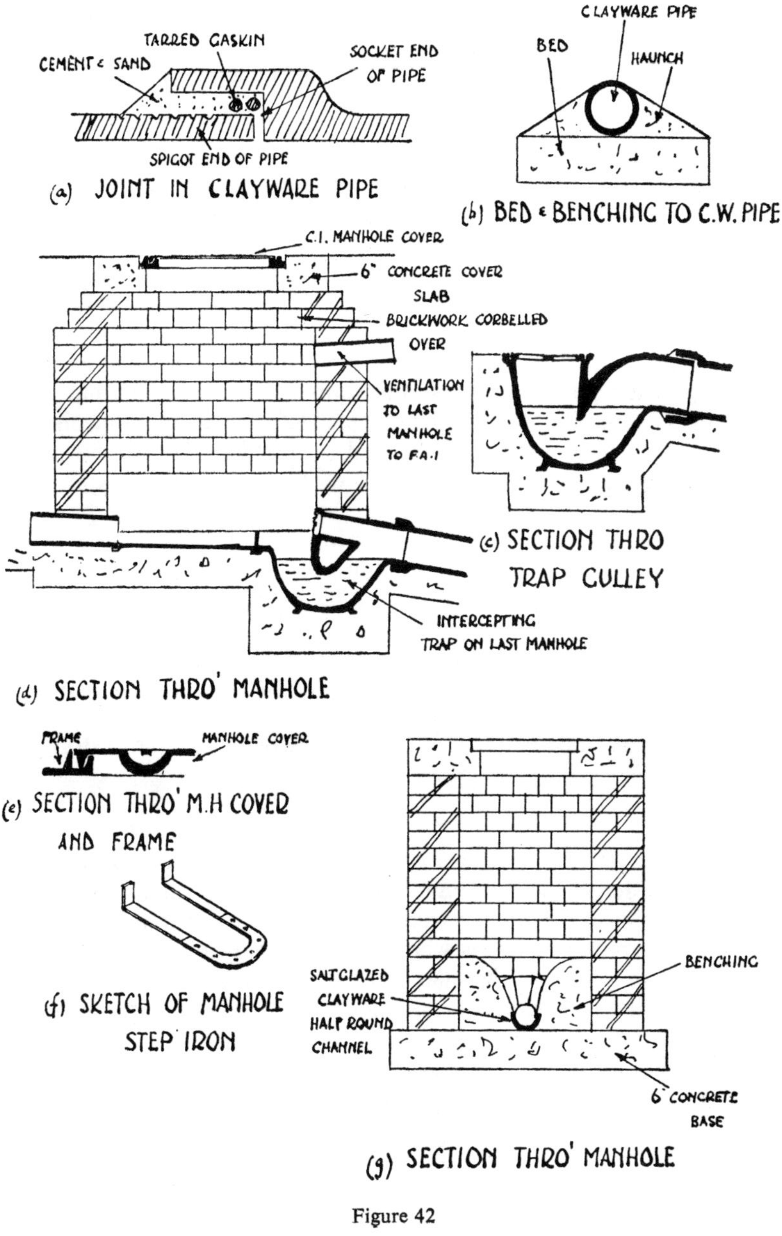
TARRED GASKIN
CEMENT & SAND
SOCKET END OF PIPE
SPIGOT END OF PIPE
(a) JOINT IN CLAYWARE PIPE
CLAYWARE PIPE
BED
HAUNCH
(b) BED & BENCHING TO C.W. PIPE
C.I. MANHOLE COVER
6" CONCRETE COVER SLAB
BRICKWORK CORBELLED OVER
VENTILATION TO LAST MANHOLE TO F.A.I
(c) SECTION THRO TRAP GULLEY
INTERCEPTING TRAP ON LAST MANHOLE
(d) SECTION THRO' MANHOLE
FRAME
MANHOLE COVER
(e) SECTION THRO' M.H COVER AND FRAME
(f) SKETCH OF MANHOLE STEP IRON
SALT GLAZED CLAYWARE HALF ROUND CHANNEL
BENCHING
6" CONCRETE BASE
(g) SECTION THRO' MANHOLE

Figure 42

6. Ordinary Quality pipes. Sound pipes not claiming to conform to the British Standard Specification.
7. Second Quality pipes. Pipes having glazing flaws, being suitable for storm-water drainage.

Saltglazed clayware drainpipes are jointed with cement and sand in the proportion of 1 : 1, a neat collar being formed around the socket of the pipe. To prevent the cement and sand penetrating to the inside of the pipe gaskin, which should preferably be tarred, is placed in the pipe joint as shown in Figure 42(a). Drainpipes are usually 2 ft long.

Since saltglazed clayware pipes will crack under excessive load, it is necessary to lay them on a bed of concrete (1:6) six inches wider on each side of the pipe and four inches thick. The pipe is held in position and protected on the concrete bed by haunching, i.e. concrete of similar mix as the bed is placed at the sides of the pipe in a triangular form and finished off with a spade as shown in Figure 42(b). Should clayware pipes pass under buildings or paths, etc., they should be surrounded with concrete.

Drainpipes are provided with fittings, the main fittings being:

Junctions which can be single or double providing for one or two branches respectively.

Bends. A variety of bends are available from 'knuckle bends' which are very sharp bends to 'slow bends' which provide a very wide radius.

Tapers. Pipes tapered in their length to increase the size of drainpipe.

Manholes

Manholes are provided as a means of access to the drainage system in order to rod it when a blockage occurs. In domestic work it is found that it is better to restrict the number of manholes, since these often are the positions where stoppage takes place. Manholes should be provided at the change of direction of the main drain and in positions where the system crosses the boundary of the property to enter the sewer.

Manholes will consist of a concrete base 4″ thick with 9″ brick sides and reinforced concrete cover slab 4″ thick, having an access hole formed in it. See Figures 41 and 42.

The bricks for the manhole should be of engineering type, or where common bricks are used, these should be rendered, preferably externally, with cement and sand. The base of the manhole will have a saltglazed clayware channel laid on it of the same diameter as the pipe discharging into it. Should the outgoing pipe be of a larger diameter, the channel may be of taper section, or the outgoing pipe be provided with a taper pipe.

The bottom of the manhole will be benched with cement and sand to the sides of the channel, the purpose of this being to direct any surplus water which may splash over the channel, back into the system.

Where other lengths of pipes or branches discharge into the manhole, channel branches will be used. The type of channel branch will depend upon the angle at which it enters the manhole, since all branch drains must discharge into the main drain in the direction of its flow. Thus where a drain length meets a manhole, as shown in Figure 41(b) the type of channel branch will be as indicated.

The last manhole in a system of drainage, before entering the sewer, is known as an *inspection chamber* and is provided with an intercepting trap as shown in Figure 42(d), to prevent the sewer gas coming into the private drain. It has been found that this fitting after a period of time loses its seal in the rodding arm and becomes ineffective, and a source of stopping in the system. It has been largely eliminated in modern systems.

The access to manholes will be sealed by the provision of a cast iron manhole cover and frame. The frame will be set in the concrete cover slab and the cast iron cover will be set in grease into the frame as shown in Figure 42(e). The usual size of manhole covers for domestic work will be 18″ × 18″ or 24″ × 18″. Where the manhole is too deep for easy access, manhole step irons will be built into the sides as shown in Figure 42(f). These will be of galvanised iron and often of 'U' pattern, which will allow a hand hold as well as a foot hold when in use.

The size of manholes will be governed by the number of branches and the depth. For shallow manholes in domestic work a common size is 2′ 0″ × 1′ 6″.

Since the building being considered is a bungalow and the outlet from the w.c. pan discharges directly into the drainpipe as shown in Figure 40(g), a soil pipe and ventilation of the pan is not required. The drainage system will, however, require ventilation and a ventilating pipe will be connected to the highest manhole, and a fresh air inlet provided at the lowest manhole. The ventilating pipe will be of cast iron or asbestos-cement and the jointing of these pipes is shown in Figures 40(e) and (f).

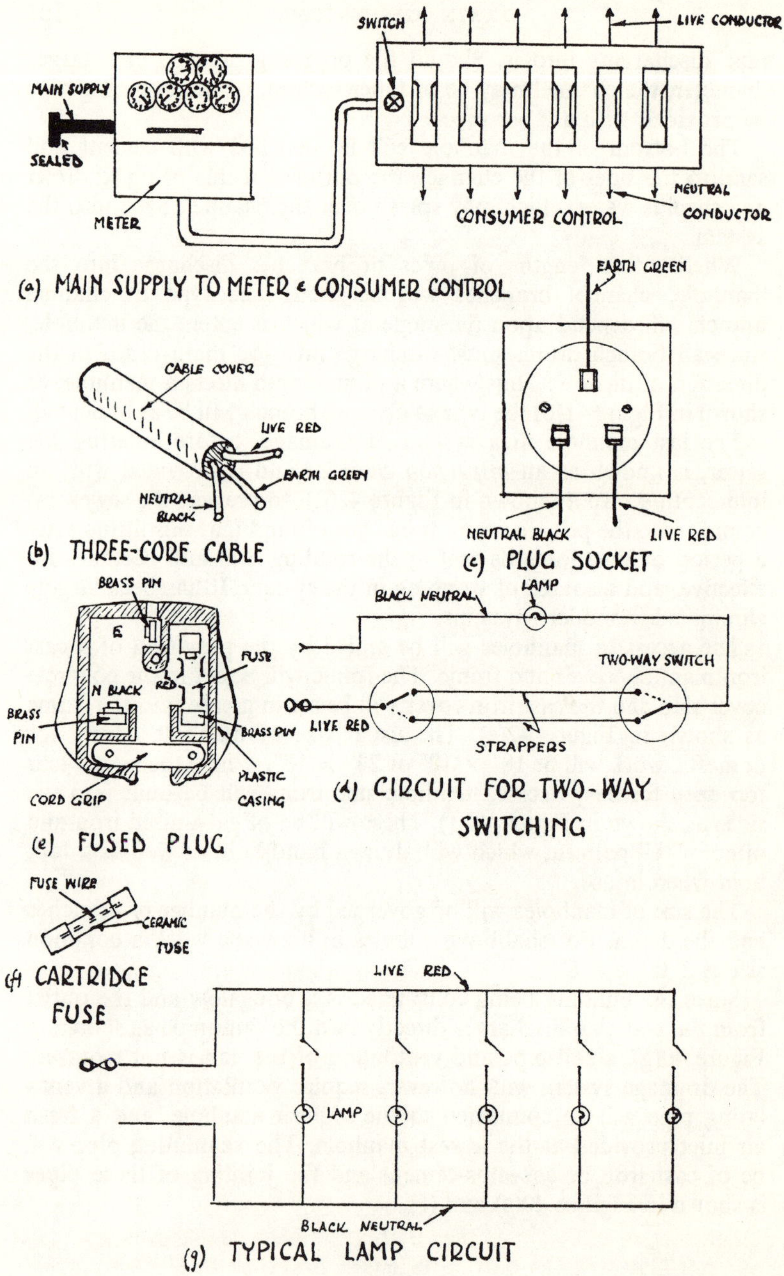
SWITCH
LIVE CONDUCTOR
MAIN SUPPLY
SEALED
METER
NEUTRAL CONDUCTOR
CONSUMER CONTROL
(a) MAIN SUPPLY TO METER & CONSUMER CONTROL
EARTH GREEN
CABLE COVER
LIVE RED
EARTH GREEN
NEUTRAL BLACK
(b) THREE-CORE CABLE
NEUTRAL BLACK
LIVE RED
(c) PLUG SOCKET
BRASS PIN
E
FUSE
N BLACK
RED
BRASS PIN
BRASS PIN
PLASTIC CASING
CORD GRIP
(e) FUSED PLUG
LAMP
BLACK NEUTRAL
TWO-WAY SWITCH
LIVE RED
STRAPPERS
(d) CIRCUIT FOR TWO-WAY SWITCHING
FUSE WIRE
CERAMIC TUSE
(f) CARTRIDGE FUSE
LIVE RED
LAMP
BLACK NEUTRAL
(g) TYPICAL LAMP CIRCUIT
Figure 43

16 Electricity and Gas Services

Electricity Service

THE provision of electricity to the building will be by the Electricity Board and will terminate in a meter upon its entry into the building. From this point onwards electrical contractors can take over. The electrical supply for new buildings will be provided on a ring main and will be three phase and 240 volts. All appliances and lighting will be fed from this main and it is advisable to have a sufficient number of points available rather than to supply appliances from one point, by means of adaptors.

The entry of the supply to the building will normally be underground from the main supply outside the boundary. The depth of the trench should be sufficient so that the cable is not disturbed by work on the ground above. A duct should be provided by the contractor through, or under, the foundations and across the building to a point below the meter and then vertically to approximately ground level. This duct will allow the electric cable to be passed inside the building adequately protected yet accessible at any future date. The meter will be placed in some convenient position, e.g. in a meter cupboard in the kitchen, garage or outbuilding accessible for reading easily. The meter will be sealed by the Electricity Board and the contractor will provide a consumer control box immediately adjacent to it. The purpose of the box is to isolate circuits of the building and so restrict the overall electrical load at any one point. The fuse will limit the loading in its circuit by providing a weaker link than the wiring itself thus ensuring, if overloading takes place, that this will fail first. Care should always be taken, therefore, that fuses are not fitted of greater strength than that stipulated and that at all times correct fuse wire is used and not any makeshift metal wire that is at hand. See Figure 43.

The electrical system should be earthed. This is often provided by linking the earth wire to a cold water pipe, but care must be taken that the cold water pipe is the rising main, that it is of metal and not plastic in any part of its length, and that the electric current

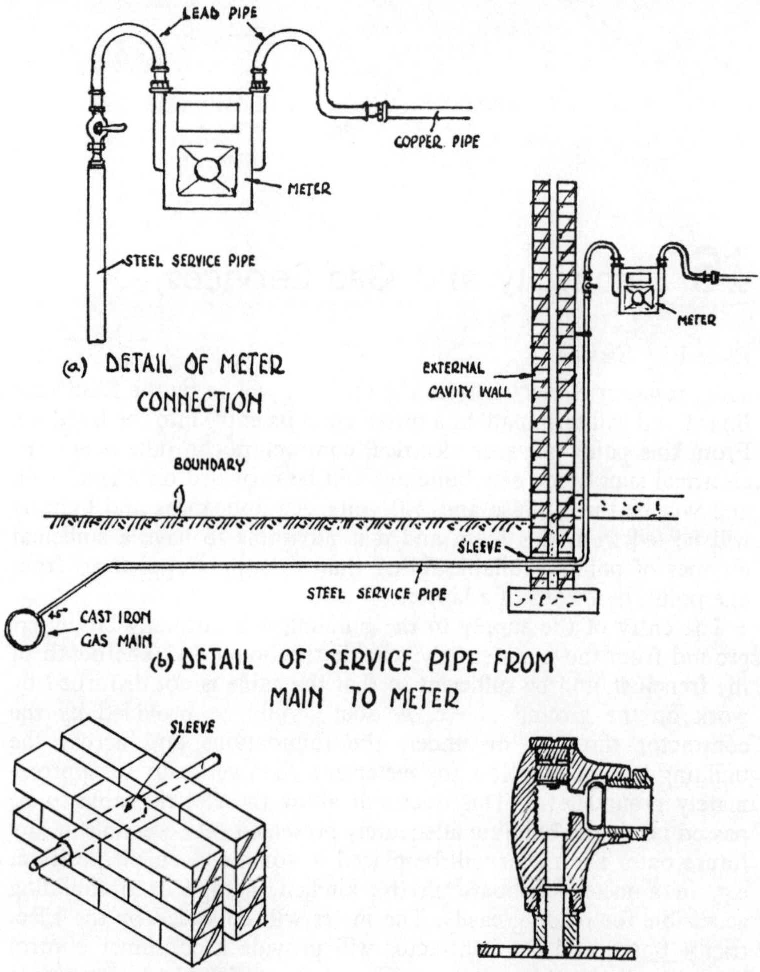
LEAD PIPE
COPPER PIPE
METER
STEEL SERVICE PIPE
(a) DETAIL OF METER CONNECTION
METER
EXTERNAL CAVITY WALL
BOUNDARY
SLEEVE
STEEL SERVICE PIPE
45°
CAST IRON GAS MAIN
(b) DETAIL OF SERVICE PIPE FROM MAIN TO METER
SLEEVE
(c) SKETCH OF SLEEVE
(d) FLOOR COCK
(e) SKETCH OF JIG FITTING
(f) CONSUMER CONTROL

Figure 44

is not allowed to escape through the pipe. The latter will be dangerous for the occupants and also electrolysis may be set up in the pipe.

The wiring from the meter will be carried around the building within a chase in the walls and above the ceiling. Metal ducting should be provided for the buried cables to isolate them and also prevent accidental puncturing of the cable by nails, etc., after plastering.

The supply cables in the roof space may be laid over the ceiling joists. Should it be necessary to provide holes in the joists, this should be at the centre and not at the top or bottom edge, since at these extreme positions the joists will be weakened.

The supply points to rooms will be provided by power plugs and light fittings as given in Figure 43, the choice of these being the client's or architect's.

An outline of a typical lighting circuit is given in Figure 43(g).

Where two-way switching is required for bedrooms, garages, etc., the circuit will be as given in Figure 43(d).

Power circuits in the older house were separated from the lighting circuits, but in new buildings the one circuit is used for both.

Gas Service

The service will be connected to the cast iron main outside the boundary of the site. The pipe will leave at 45° to the vertical cross-section of the main and will be laid in a trench up to the building where it will enter at right angles to the wall face. The service pipe will be 1″ diameter mild steel (Class C) pipe with screwed joints, pre-wrapped in hessian and bitumen. The depth of the main will be 2′ 3″ below ground level and the service pipe should be at least 10″ below ground level and fall towards the main at a gradient of 1 in 120.

Where the service pipe has to pass through a concrete floor a duct should be provided at least 6″ greater than the pipe.

The meter must be fixed in an accessible position for reading and in the case of pre-payment meters for the insertion of coins. The inlet and outlet connections to the meter will be in lead with solder joints. A consumers' control cock will be provided on the inlet to the meter. The meter for the bungalow will be an I.G.E. (Institute of Gas Engineers) meter Type D2, 200 cu. ft per hour capacity. See Figure 44.

The service pipe from the meter will be $\frac{3}{4}$″ diameter copper tubing with capillary joints to a point beyond the cooker, after which it may be reduced to $\frac{1}{2}$″ diameter tubing to service the gas fires. Should an instantaneous water heater be required at the sink a 'jig' fitting may be provided with flush plug connection, as shown in Figure 44(e).

Where pipes pass through walls sleeves will be provided and where they are laid in concrete floors a p.v.c. cover should be provided to protect the pipe.

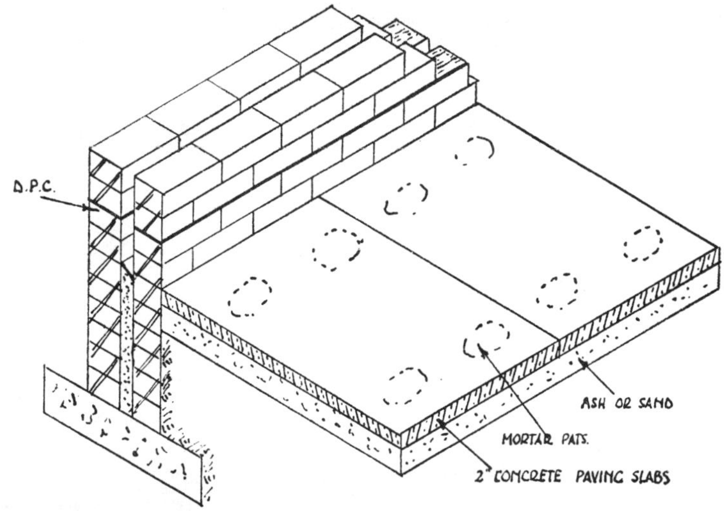

(a) PRECAST CONCRETE PAVING

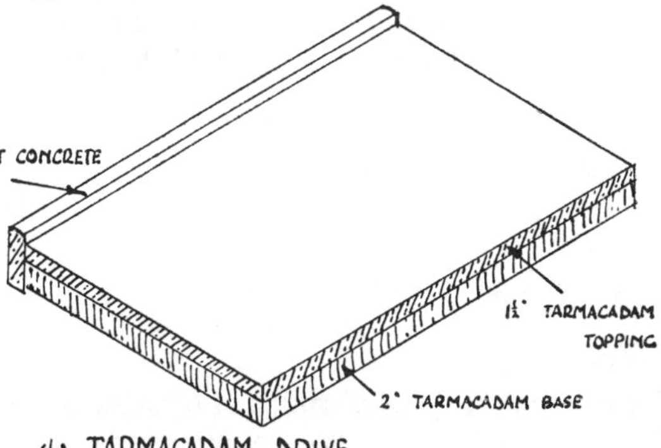

(b) TARMACADAM DRIVE

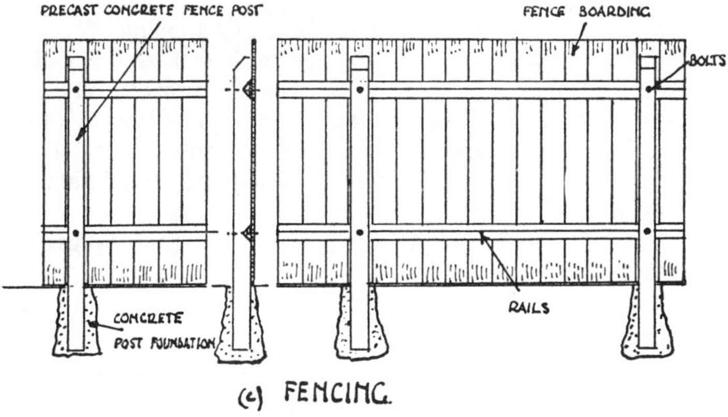

(c) FENCING.

Figure 45

17 External Works

Paving

PATHS around the building should be paved to allow clean access. The top soil must be removed from the area to be paved and the subsoil excavated to a firm bottom. Ashes or sand 3″ consolidated thickness should be laid on the subsoil and the paving slabs laid on this. The ashes or sand will provide an even bed for the paving and on this can be laid cement mortar, usually at the corners to fix the paving. After laying, the joints should be pointed, often carried out by grouting, but this latter method can spoil the appearance of the slabs.

Paving slabs are manufactured under factory conditions, by pressure or vibration and should be adequately cured before delivery or laying. The size of slabs will be 2′ 0″ × 2′ 0″ and 3′ 0″ × 2′ 0″, 1½″, 2″ or 2½″ thick.

Typical path laying is shown in Figure 45(a).

Drives to Garages

The finish to drives to garages can be of concrete or tarmacadam. Tarmacadam is a mixture of small limestone and tar or bitumen. The top soil should be removed over the area of the drive and the subsoil excavated to accommodate the base and topping. The base may consist of small hardcore, if the subsoil is soft, well rolled to form a firm bed. Ashes or sand can be rolled into this to fill up all gaps in the hardcore. A layer of coarse tarmac is laid on this base and well rolled with a hand roller to give a consolidated thickness of 2½″–3″. A further layer of fine tarmac is laid and consolidated as before to give a thickness of 1½″. This will give a fine finish to the drive and be sufficiently strong to take car traffic from the garage. See Figure 45(b).

Fencing

Fence posts should be of oak or other durable hardwood. They should be set in concrete after excavating a hole of sufficient depth to give a firm fence. Timber rails will be fixed to the posts and boarding fixed to the rails. See Figure 45(c). All timber used in fencing should be treated with a preservative, preferably one of the preservative treatments under pressure.

Galvanised wire mesh may be fixed to metal posts to give a cheaper and more open type of fencing.

Index